The Real Chimpanzee

Sex Strategies in the Forest

The Real Chimpanzee encapsulates the fascinating behaviour of wild chimpanzees and discusses the differences observed in different populations across the species, and across the many levels of their social behaviour. It explains why sex competition and predation pressures in a forest chimpanzee population made the females of the group highly social and gave the males a high level of within-group solidarity, making them very xenophobic towards outsiders. Love is what makes war possible. Christophe Boesch brings back to the table the debate over ecological pressures and social organization, and the influence they have over issues such as the evolution of warfare, cooperation, altruism and the position of females. Written in an accessible style for a general audience, as well as for undergraduate and graduate students, he presents insightful views to give readers the background information to understand the struggle for survival of our closest living relative, the chimpanzee, and through this to find some keys to the ever-so-intriguing question of what makes us human.

CHRISTOPHE BOESCH is Professor and Director of the Department of Primatology at the Max-Planck Institute of Evolutionary Anthropology in Germany.

The Real Chimpanzee

Sex Strategies in the Forest

CHRISTOPHE BOESCH
Max-Planck Institute of
Evolutionary Anthropology

CAMBRIDGE UNIVERSITY PRESS
Cambridge, New York, Melbourne, Madrid, Cape Town, Singapore, São Paulo, Delhi

Cambridge University Press
The Edinburgh Building, Cambridge CB2 8RU, UK

Published in the United States of America by
Cambridge University Press, New York

www.cambridge.org
Information on this title: www.cambridge.org/9780521110082

First published 2009

Printed in the United Kingdom at the University Press, Cambridge

A catalogue record for this publication is available from the British Library

ISBN 978-0-521-11008-2 hardback
ISBN 978-0-521-12513-0 paperback

To my cousins in the forest, *Brutus, Kendo, Ondine, Loukoum*
Who were so patient with me
To my kids, *Lukas, Léonore*
Who lived through my passion of the forest
To my father, *Ernst Boesch*
Who enthusiastically challenged my interpretations
To my wife, *Hedwige*
Who made it all possible

Contents

Acknowledgements   *page* ix

1   **Make love and war?**   1
    From the dream of the hippies to the reality where sex
    and violence often intrinsically intermingle to produce
    some of the complex strategies used by both sexes to
    find partners and reproduce.

2   **Inconspicuous female superiority**   11
    How the traditional image of females as the passive and
    submissive sex has changed into the sex determining
    and controlling reproduction to her advantage in the
    competition with more powerful males, and how
    a female manipulates males to select the best sperm
    for her offspring.

3   **The tyranny of the testis**   33
    How males have been able to develop, within a system
    of dominance and aggression, close tight cooperative
    units that fight for the good of group members and
    how this has made altruism become one of the most
    impressive behaviors contributing to the survival
    of individuals.

4   **Odyssey through our forest past**   60
    Sex and cooperation in the forest: what is life like as
    a large social primate in a dense tropical forest where
    the visibility is restricted to 20 metres and where most
    aspects of sociality have to be inferred and communicated
    by vocalizations?

5   **Make war to get love**   76
From the individualistic struggle of both sexes to find
suitable partners and resources to cooperative teams
increasing reproduction and how that leads to both fatal
violence and altruism, while females pursue sexual
exchange that males cannot prevent.

6   **The real chimpanzee**   109
From a millennia-long past in Central African forests to
adaptation into more open habitat regions in East Africa
as well as in savannah-like regions – chimpanzees have
conquered many regions and this expansion into more
marginal habitats resulted in some dramatic shifts in
males' cooperation and altruism as well as in the females'
control over reproduction and social position.

7   **When sex becomes destructive**   138
Why humans, one of the most cooperative and altruistic
species, became so destructive throughout their range
and why some of the challenges solved peacefully by
chimpanzees have become so destructive in humans.

8   **Postscript: Fédora's fate**   160
Will Fédora, the chimpanzee baby I saw develop and
become a skilful tool user, survive the loss of her hand
to poachers and be able to survive as a young successful
mother? In remote areas, chimpanzees are tracked by
humans; without decisive and rapid action they have
no future, and our cousins, our roots to our past, will
vanish before we get to know them.

References   164
Index   174
Plate section between pp. 82 and 83

Acknowledgements

As young people, many of us struggle to find out about our destiny in life, how to develop as an adult and in which field to find an interesting profession that, hopefully, also allows us to make a living. I was lucky enough to become inspired by some books around the age of fourteen. It was *King Solomon's Ring* by Konrad Lorenz which was the enlightening experience for me. I was so impressed by his intimacy with geese and jackdaws that my path was found – I wanted to work with or on animals. Some years later, I discovered *The Year of the Gorilla* by George Schaller, and then I simply knew! Incredibly enough, a few years later, as I had finished my biology studies, I was able to follow the same gorillas George Schaller had so much adored.

Observing animals is more than a passion, it is a profession. For me, the young Swiss scientist, the field of primatology was an irresistible attraction, much more so than any other one in biology. I was lucky enough to find mentors who made my reach into this exotic scientific field possible. It all started on a good track as one of my professors at the University of Geneva, Arianne Etienne, knew the late Dian Fossey personally, and she made it possible for me, in 1973, to start my first real experience of wild primates at the top with no less than the mountain gorillas, there in one of the pinnacles of nature, in the unforgettable world of the high mountains of Rwanda and eastern Congo. Following this experience, I had another stroke of luck. In Paris, I met the late French professor François Bourlière who mentioned some reports about remains of nut pounding in the forest block of Liberia and Côte d'Ivoire, entertaining the so far unconfirmed possibility that chimpanzees there might be using tools to crack nuts. Supported by the professors Hans-Jörg Hugel and Rudolf Schenkel from Geneva and Basel universities respectively, I was able to go to the Taï forest and, after many months of searching, could confirm with one single observation that, indeed, it was the chimpanzees who were pounding these nuts. This one observation made my future as a primatologist possible.

In 1977, after this initiation into the world of chimpanzees in the African rainforest, Hans Kummer, professor at the University of Zürich, accepted the risk of supporting an unknown student, initiated me to the science of primate research, was my PhD director and then unfailingly supported me as we initiated the Taï chimpanzee project and went through the many years of habituating the chimpanzees to human presence, during which process scientific advances are often frustratingly slow. Hans Kummer's intellectual mix of rigour and keen striving for the unknown have marked my own attitude in science. My present involvement in primate research and conservation is part of the legacy he left me.

In 1991, Steven Stearns, professor at the University of Basel, after visiting the Taï chimpanzee project, offered me the possibility of an academic position at his institute while being able to continue the field study in the Taï forest. This generous offer allowed me after 12 years in the African forest to touch base with scientific progress and to intensify the Taï chimpanzee project as well as to develop my scientific career by being based in Switzerland. I am immensely grateful to Steve for this chance to participate in his challenging and stimulating research group.

I also wish to express here my deepest gratitude to Professor André Aeschlimann from the University of Neuchâtel, head of the Swiss National Foundation and President of the CSRS in Abidjan for many years. He supported the Taï chimpanzee project throughout the important early years and therefore contributed to its achievement as a long-time study now in its thirtieth year.

My wife Hedwige Boesch-Achermann was enthusiastic, curious and open to leaving the so-called civilized world. She accompanied me from the very first day and for more than 12 years of permanent life in the Taï forest. If the study of the Taï chimpanzees was successful, it is only because she was there to live and work with me during all these years. This transformed our life into an adventure, a long inquiry and discovery of some of the most intimate aspects of the life of our cousins from the forest. Most of the ideas developed in this book have matured over the years we spent together in the forest, but also afterwards, when she continued her work for the chimpanzees by staying in Europe. Trying to understand this elusive species and discovering the many aspects of their life was and is part of our life. Therefore, throughout the text, the use of 'I' should often be read as 'we' as the Taï experience was a joint experience which moulded so many aspects of my scientific and personal way of thinking.

From the start we looked at our project as the new one compared with the famous long-time research projects of Jane Goodall and her colleagues on the chimpanzees of Gombe Stream National Park and of Toshisada Nishida and colleagues on the chimpanzees of the Mahale Mountains National Park, both in Tanzania. It was therefore a great honour to us when both Jane Goodall and Toshisada Nishida invited us to their study sites and encouraged us to compare our results from the Taï chimpanzees. Not only was this a wonderful opportunity for us but it was an eye-opener to what population differences in this species really means. I am, therefore, not only very grateful to both pioneers of chimpanzee research to have invited and supported me, but equally to Frodo, Fifi, Wilkie and Co., the chimpanzees from Gombe, as well as to Fanana, Alufo, Gwekulo and Co., the chimpanzees from Mahale to have taught me so generously about how chimpanzees behave in their distinct social physical world. Once the fascination about chimpanzee behavioural diversity was seeded in me, I was very lucky to be able to enjoy the generous attitude of colleagues who invited me to visit their sites and observe their study chimpanzees to continue my comparison on the behaviour of the chimpanzees. I am thus very grateful also to John Mitani and David Watts and the Ngogo chimpanzees, to Vernon Reynolds and the Budongo chimpanzees, to Crickette Sanz and David Morgan and the Goualougo chimpanzees, and to Richard Wrangham and the Kanyawara chimpanzees, for accepting me at their sites and for guiding my vision into the fascinating individuals of these study communities.

As the Taï chimpanzee project developed, we have started to include both students and local field assistants to increase the number of observations to be done with the chimpanzees as well as to habituate three more new chimpanzee communities. This has made the Taï chimpanzee project the first one to study neighbours and their interactions. All of the many involved field assistants have our deep gratitude. I wish to single out a few. Grégoire Kohou Nohon was the first local assistant who joined us, at a time when most people close to Taï National Park considered the forest as dangerous. Grégoire overcame his natural fear and said right from the beginning that he would stay on. And so he did. After over 20 years, he is still working with us! Grégoire's example impressed many people in the region and thanks to him we have been able to find many more young men who all became dedicated chimpanzee observers. Not only were they willing to spend 12 to 14 hours per day following the chimpanzees on their forays in the forest, often over 5 kilometres long, but they were spending nights out in the forest when there was a need to protect seriously ill chimpanzees from possible leopard attacks. And they were ready to keep going during the very unstable period of the civil war from 2002 to 2004, thereby actively protecting the study site and the chimpanzees. They have been our ambassadors in the local populations and made the development and continuation of the project possible during the 30 years since its start in 1979. We especially thank the senior field assistants Kpazahi Honora, Bolé Camille, Oulaï Nicaise Daurid, Bally Louis Bernard, Tahou Mompeho Jonas, Gouyan Bah Nestor, Gnahe Djirian Appolinaire, Yagnon Valentin, Tah Alain Pahi, Sioblo Arsène, Guy Sylvain, Guiro Ferdinand Thia, Kevin Charles Bally, Blaise Blé Téré, Ignace Dezaï, Camille Dji, Gabriel Gnombouhou Kouya, Gérald Gah, Benjamin Goullaon, Mathias Douosson, and Denis Lia.

At the same time, I thank the many students, far too numerous to be named individually, from all over the world who helped to follow and study the chimpanzees in the Taï forest for short periods. I thank all students and colleagues who were involved in projects for longer periods for their dedication and passion for the chimpanzees and their forest: Barbara Fruth, Paul and Nathalie Marchesi, Frédéric Joulian, Margaret Hoitink, Pascal Gagneux, Miriam Behrens, Gerd Radl, Christian Falquet, Diane Doran, Martina Funk, Rainer Neumeier, Andy Kurt, Penny Simpson, Brigitte Schmid, Suzanne Pieren, Ulrike Ratkjen, Annemarie Fränkl, Gregory Roduit, Paco Bertolani, Chloe Cipoletta, Dean Anderson, Nick Malone, Ilka Herbinger, Roman Wittig, Myriam Sele, Cathy Crockford, Steven Bada, Nicola Paterson, Daniel Hanus, Hjalmar Kuehl, Julia Riedel, Janna Rist, Melissa Tauber, Kathleen Beese, Sabrina Locatelli, Lionel Egger, Pola Abaza, Sandra Junglen, Rebecca Stumpf, Antoine N'Guessan, Zoro Goné Bi, Anja Blankenburg, Célestin Kouakou, Simone Ban Dagui, Nadine Eckhardt, Lydia Luncz, Livia Wittiger, Sonja Metzger, Svenja Schenk, Siva Aina Jensen, and Fabian Leendertz.

In any long-term projects, some difficult periods may occur that might have dramatic consequences for the chimpanzees we habituated to human presence. Thanks to the courage and altruism of some individuals, the worst – the disappearance for unknown reasons and/or the killing of the chimpanzees – was mainly prevented. My warmest thanks therefore to Claudia Steiner and Franca Donati who were present during the emotionally hard period of the Ebola outbreaks in the Taï chimpanzees in 1992 and 1994, to Thomas Pfluger who stayed near the camp when, in 1991, unrest from Liberia spilled over into the Taï region, and to Emmanuelle Normand, Yasmin Moebius, Tobias Deschner and Cristina Gomes for having decisively contributed to keep the project going during some of the most difficult times of the civil unrest in Côte d'Ivoire during 2002 to 2004.

I am most grateful to Valerie Howe for her hard work with the language correction of this book. And I thank Hedwige Boesch, Tobias Deschner, Katerina Guschanski, Lydia Luncz, Kevin Langergraber and Heike Siedel for comments on an earlier draft.

Our long-term study in the Taï National Park was possible only with the constant and amicable support of the Ivorian authorities. Thirty years is a long period and by far exceeds the normal life of any government; nevertheless the support from the Ivorian government remained inflexible. We wholeheartedly admire the constancy of the support we received from all the different persons working in and heading the 'Ministère des Eaux et Forêts', as well as within the 'Ministère de la Recherche Scientifique', in particular those working and heading the 'Direction de la Protection de la Faune', of the 'Office Ivoirien des Parcs et Réserves', and of the 'Direction du Parc National de Taï' as well as the numerous park agents, especially those of the Taï Sous-Préfecture. The Taï National Park has been throughout our study subject to attacks from various directions, mainly in the shape of logging, poaching, civil unrest and farming, and this has threatened the survival of the forest and its fauna, including the chimpanzees. The Ivorian authorities have always taken the steps necessary to guarantee the survival of this precious park, and unfailingly supported the continuity of the project. The Centre Suisse de Recherches Scientifiques en Côte d'Ivoire (CSRS) has been our base in the capital Abidjan since our very first visit to the country in 1976, and they have always remained a key partner in the project. We thank all the successive directors and presidents for going out of their way to support us. In Taï, we profited a lot from the support of the directors of the Station d'Ecologie Tropicale and their staff, Denis Vivet, the late Théo Tiépkan Zoroa and Paul Zouhou respectively. Special thanks also go to my colleagues, professors and dean, of the University of Cocody and Abobo from Abidjan for supporting the collaboration between our institutions.

No scientific project can survive without financial support, and here we have to stress the continuous, generous and unvarying support we received for 19 years from the Swiss National Science Foundation and since 1999 from the Max Planck Society. Very few public funding organizations dare to support a long-term field project for so long, and we are immensely grateful for their appreciation of such a study. We are grateful for additional funds from the Messerli Foundation, the Leakey Foundation, the Schultz Stiftung, the Jane Goodall Institute, the Wenner Gren Foundation, the Roche Foundation and the Freie Akademische Gesellschaft.

How dull would our life have been in the forest without all the encounters with the wonderful, friendly, helpful, curious and tolerant African people near and in our forest camp, and their joy in sharing their knowledge with us? What would our life have been throughout all these years without the friends and family from abroad who have visited, sent presents, letters, food, books, and stayed in contact while we were in Africa and back in Europe? Thank you, to all of you.

Last but not least, we thank all the chimpanzees who have tolerated our presence for so many years, allowed us to share so much of their life with us and made us wonder from day to day about our own species. In the process of habituation and following them we have at the same time inadvertently contributed to their confrontation with threats, such as diseases, and made them more vulnerable to poachers. This has been a constant worry for us from the beginning and I hope that this book, by increasing the standing of the chimpanzees in the scientific community and the public at large, will allow us to pay back some of the debts we owe them.

1

Make love and war?

Everyone is familiar with the hippy mantra 'make love, not war' – an ideal that inspired a generation. Looking at the world and our past, however, it is 'make love *and* war' rather than the 'flower power' ideal that seems to be the normal condition. Philosophers, such as Jean Jacques Rousseau and Thomas Hobbes, proposed that war is a natural state for humans resulting from the increasing competition in the growing human population following the advent of agriculture. Nowadays 'make love and war' seems to be the natural state for humans, whether it be in highly technically advanced and 'sophisticated' societies, such as those involved in the major regional wars of the twentieth century, or in less technically developed societies, like traditional hunter-gatherers, such as the Nuers and the Nimba of Sudan, the Australian Aborigines and the Batak of Asia or the Jivaro and the Yanomamö Indians of South America. Some traditional societies are constantly at war, while others have periods of relative peace, but it was shown that a state of war is characteristic of nearly 95% of all known human societies. When in our past did this situation start to prevail? What are the benefits of war? How did it happen that humans are so violent when at the same time showing so many cooperative and altruistic propensities? Do we see the same coexistence in chimpanzees? How can we learn from them to understand us?

I propose that by looking at our closest living relatives, the chimpanzees, our cousins of the forest, we can gain insight as to why altruism and aggression coexist in the same species. Knowledge acquired from other animal species can help us to improve the understanding of our own nature. The crucial point here is that in all communities of wild chimpanzees studied we saw from time to time extremely violent interactions with neighbours, and such conflicts can result in individuals being killed and groups annihilated. Chimpanzees are known, with humans, for being the only primates able to make and use tools as well as to hunt for meat in groups. Warfare provides another similarity between humans and chimpanzees.

The similarities between chimpanzees and humans in these behaviours have been highlighted by previous authors, in such famous books as *The Naked Ape*, *Chimpanzee Politics*, *In the Shadow of Man* or *Demonic Males*. Classically, the origin of war has been proposed to result from an inbuilt drive for aggression in both human and chimpanzee males that produce high levels of violence. Male bonding was proposed to result from this common tendency for violence among males on which chimpanzees' ability to form alliances would be overlain. Arguments about humans' appetite for violent behaviour are presented that set humanity apart from the violence shown by other animal species. Humans' ability to incorporate tools into social conflicts would enhance the potential for this destructive and 'unnatural' aggression.

LOVE IS WHAT MAKES WAR POSSIBLE

I came to this very different conclusion after years of following the forest chimpanzees. At a first glance this may sound contradictory, but if you follow my reasoning, you will acknowledge, I hope, that there is some intuitive sense to such a proposition. To make war and to respect someone for killing other human beings, you must first have an extremely strong sense of belonging to 'one group', which leads to a disregard for the human dimension of outsiders. It is this 'dehumanization' of non-group members that makes war at the same time both special and disturbing. To me, this is the natural consequence once cooperation and altruism with, and towards, other group members have infiltrated every aspect of social life. Only then will outsiders be despised to such an extent that killing them will not only be acceptable but even sometimes applauded. Xenophobia, the hatred of outsiders, results from extended within-group solidarity. Thus, in contradiction to the hypotheses which propose that war is caused by destructive forces, I suggest that war results from strong within-group solidarity.

What makes solidarity within groups so important? The year-long observations of the chimpanzees of the Taï National Park in Côte d'Ivoire revealed how predation pressure forces individuals to seek protection among other group members and results in both sexes spending more time together than they would in the absence of predators. The more time one spends with other individuals, the less likely one is going to be singled out in a leopard attack. Once individuals gathered to protect themselves from leopard attacks, two social innovations followed. First, the more individuals spend time together, the more they are going to compete with one another for access to food resources. This exemplifies the major dilemma of group living, as the cost of competition for food is inherent and increases with the number of individuals together. This could be alleviated by developing special relations with specific individuals through increased coalitions and friendships, so that such partners support one another in social conflicts. Second, cooperation and altruism between group members represent the best way to counter the negative effects of attacks by powerful predators such as leopards. Single individuals have hardly a chance when facing a leopard, while cooperation and altruism can be life-saving. Once established, the ability to cooperate can be extended to other aspects of social life and that is what we see in Taï chimpanzees.

In addition, in chimpanzees, within-group solidarity coincides with between-group hostility. The main reason is that females in long-lived and slow-growing animals, like humans and chimpanzees, are for the most part of their lives occupied with caring for dependent offspring, periods during which they are not fertile. Thus, adult males, even if they fight one another for access to females within the group, have a great common interest in joining together to win access to additional females from other groups. Thus, sex competition that is predominantly an individual challenge becomes a social challenge in slow-maturing species facing high predation pressure. This shift from individual to social is not limited to the sexual domain, as the pursuit of reproductive success does not restrict itself to mating. The acquisition of new mates plus stable social conditions and secure access to food resources are all needed to improve reproductive success and could be improved and secured through collective warfare. This duality between competition and cooperation is one of the key elements of male sociality, both in chimpanzees and in humans.

Both humans and chimpanzees are long-lived, slow-growing social primates that faced high predation pressure from, respectively, cave bears, lions and sabre-tooth tigers for the former and leopards or lions for the latter. Basic common biology in both species predestined them to respond similarly to high predation pressure. What I am proposing is that some salient environmental conditions experienced by chimpanzees in the rainforests of Africa are similar to those faced by our ancestors and that war resulted from such prevailing ecological and demographic pressures. It was the development of strongly affiliated and supportive behaviours within the social group that resulted in warfare and this appeared very early in the shared human/chimpanzee evolutionary line.

TELL ME WHAT YOU WEAR, AND I WILL TELL YOU WHO YOU ARE

Such a saying captures the significance of the environment to everyone. Not only does one have to wear different types of clothes in Scandinavia compared with the Amazon or the Sahara, but such climatic differences will force one to eat different types of food, and to hunt or extract food in different ways, and the amount of food that can be gathered will influence the size of a group of individuals able to survive on such resources. The paramount importance of the environment is too easily forgotten for humans like us living in heated or air-conditioned houses. Nonetheless, for others, it sets the basic conditions under which the struggle of survival unfolds. No one can escape it, and each one has to use all his or her physical and intellectual abilities to solve the daily challenges. Simple things like finding water, a safe place to sleep or escaping predators require dramatically different solutions depending upon the ecology encountered. It is the daily challenges faced during growing up that mould the individual and everyone is the product of such experience.

The hugely successful book by Frans de Waal, *Chimpanzee Politics*, in 1982 has made the complex social life of the chimpanzee familiar to a large audience. The heroes of his book lived in the zoo of Arnhem in the Netherlands, where they had the luxury to live in a large group with three adult males, a unique situation for zoos at the time. Frans de Waal's precise descriptions rightly emphasized the complex strategies males were following to climb up the dominance ladder in the group, but to a person familiar with the life of wild chimpanzees, the situation in the zoo remained strikingly simple. For one, food is being brought to the animals twice a day and they simply have to sit and eat what is given to them, while wild chimpanzees spend about 40% of their time, day after day, looking for and processing food. Sometimes, food is difficult to obtain, involving the need to use tools or to hunt to capture it. Second, in a zoo, a group of chimpanzees spends all their lives together within one enclosure, while all their wild counterparts live in fission–fusion groups where rarely more than a third of the group members are seen together at a time. This requires much greater flexibility and planning when it comes to social fights and coalitions compared with those detailed in *Chimpanzee Politics*. Thirdly, chimpanzees in a zoo are protected within an enclosure and face no external danger. This is hugely different to the situation of wild chimpanzees that are constantly under the threat of being attacked by neighbouring chimpanzees and by natural predators, such as leopards or lions. Such differences have wide-ranging consequences on the behaviour of the chimpanzees, and, in many ways, to me, sometimes captive chimpanzees seem to belong to another species than the wild ones. To make sense out of the complex social behaviour we see in chimpanzees, it is essential to consider the ecological contexts under which they emerged and understand how flexible chimpanzees can react to differences in these contexts.

On the wild side of the species, the best-seller book *In the Shadow of Man* by Jane Goodall in 1970 made many people aware that chimpanzees are indeed wild animals, not only living in zoos, or to be seen riding a bicycle or dressed up as 'humans' in shameful advertisements for whatever commercials. Her book made the public attentive that chimpanzees are individuals with strong personalities, that so-called uniquely human tool skills are daily occurrences to them and that they overcome daily challenges to survive. The chimpanzees she observed live in the Gombe National Park near Lake Tanganyika, a mosaic of gallery forest with many open woodlands and savannah with a breathtaking view on the lake and the sunsets. They are part of a small population of chimpanzees constituting the easternmost distribution of the species in Africa, while the vast majority of chimpanzees have always lived in the heart of the dense tropical rainforest of Central and West Africa. This point is of importance as, for one, the amount of food available to the chimpanzees is limited in such a mosaic open habitat, while the dense rainforests of Africa with tall emergent trees loaded with fruits are providing plenty of food to specialized fruit eaters like the chimpanzees. This has important social consequences, as in rich forests, opposed to a savannah, larger groups of individuals can forage together and many aspects of the social life of the chimpanzees are influenced by such an environmental difference. Second, the higher biodiversity of the African rainforest comes not only with more fruits but with many more animal species, including large populations of leopards that are, besides man, the main predators of chimpanzees. The absence of predation pressures in the chimpanzees studied by Jane Goodall and the relative openness of their habitat seems to have led to a whole cascade of behavioural differences that we are only now uncovering and these are the central tenet of my arguments.

Yes, after over 40 years of observing chimpanzees, we are only now starting to be aware of the extent of their behavioural flexibility and it is my intention in this book to specifically address this dimension. My experience with different chimpanzee populations living in forest regions of West and Central Africa as well as in the open woodlands of East Africa made me directly aware of how radically the ecology affects their behaviour. Not only is it fascinating to discover how an animal species can so flexibly respond to challenges, but at the same time we can identify the aspects within the ecology that influence such behavioural changes. I will use this knowledge to investigate what favours warfare between social groups, and what influences differences in the choice of sexual partners, in achieving paternity success, in investing in offspring, and in exhibiting cooperation, altruism or aggression. Due to the biological and psychological proximity between humans and chimpanzees, similar factors might have similar consequences and this approach can help us to understand the evolution of such behaviour in the human line.

In the many books discussing the origin of human social and technological uniqueness, be it from a psychological, philosophical or biological point of view, the chimpanzee population of reference has mostly been the Gombe chimpanzees. As such, humans were revealed to possess a much higher level of female sociality, more complex cooperative behaviour, and to have a more developed reliance on non-vegetable food items and tool use. Other authors have based their comparisons on captive chimpanzees. Certainly captive chimpanzees are chimpanzees, but they have spent their whole life under totally artificial conditions in human-made social settings, just as human prisoners remain human beings. However, no one would seriously consider restricting studies on humans to prisoners! Chimpanzees, just as humans, belong to one of the most adaptable primate species on this planet: chimpanzees are found in the deepest rainforests all the way into the very dry savannah regions provided there are some gallery forests. I propose here to open the door to the complex world of the chimpanzees.

HOW TIME HAS IMPROVED THE QUEST
FOR OUR ROOTS

The development of field work on wild chimpanzees and other apes for the last 40 years has brought us to the point where we can start to understand 'what makes us human?', and some answers about 'what is a chimpanzee' are emerging. All traditional human societies have myths that explain the origins of humans and their relationships with other animals. Religion and science have zealously taken up the challenge and generated their own propositions. For centuries, philosophers have proposed answers to the question based on incomplete and often romantic ideas about the nature of animals.

Now the quest for 'what makes us human?' has become for the first time a science: studies on wild chimpanzees started some 45 years ago and after years of painstaking efforts, numerous, detailed, intriguing and fascinating information about eight populations throughout Africa has provided a uniquely complex and detailed, but still incomplete, picture of our cousins of the forest. Foremost, we have learned that chimpanzees are much more diverse than originally thought, to the point where it is becoming more and more arbitrary to talk about 'the chimpanzee'. We should describe them rather as the 'Taï chimpanzee' or the 'Gombe chimpanzee', just as we talk about the Inuits, the Touareg or the !Kung Bushmen in recognizing the diversity that exists in these traditional hunter-gatherer societies. Sociality of the sexes, vocalization, tool use, hunting behaviours, diet and cultural traits, all have been shown to differ profoundly between different chimpanzee populations.

Furthermore, we are unravelling for the first time some of the most secret aspects of the social life of wild animals. Biologists, philosophers and human behavioural scientists agree that one key motivation in life is the drive to reproduce and provide the best environment for the resultant offspring. If such a drive is purely instinctive and genetic in insects and fishes, it is generally agreed that in mammals, and especially primates, the fulfilment of such a drive can also be the result of individual strategic choices, depending on the specific living conditions each animal faces. Decisions might vary; for example, an individual living in a large social group during a drought in a savannah with little available food will make different decisions from an individual living in a small group in a tropical rainforest during the main fruiting season. Decisions may also differ depending on whether the individual in the first situation is a low ranking group member or a high ranking one in the second situation. Such choices are faced daily and decisions will differ over time.

What's new in this debate? The key problem we faced in the past was that fertilization in birds and mammals is internal and therefore we had no way of being sure about paternity in wild animal populations. If maternity is relatively easy to observe, paternity remained a big mystery. Now our understanding of animal sexuality has been revolutionized by the introduction of new genetic techniques in the field of wild animal studies. This silent revolution has taken place over the last two decades and is still ongoing. Being able to confirm maternity and decipher paternity of wild animals by using tiny amounts of DNA found in faeces, urine or chewed food remains changed the field and we can for the first time answer such a basic question as 'who is the father?'. To take just one example, the application of genetic methods to monogamous bird species has revealed in many species a surprisingly high percentage of paternities not attributable to the long-term male partner. What was viewed as monogamous behaviour is in many species restricted to their sociality; reproductively, both sexes can be polygamous! Much work has been carried out recently in applying these new technologies to wild chimpanzee populations. Thus, we are now in a position to compare the sexual strategies of individuals in different social groups and determine how successful such strategies are. In other words, for the first time, we can judge how evolution works at an individual level.

OUR BRAIN SHOULD SEE THROUGH OUR EYES

We all are conditioned by what we learn in our early years and this affects the way we look at things and how we interpret them. In addition, religious and philosophical considerations teach us about what the world 'really' is. To complicate matters, as students at schools and universities, we are fed ideas and theories from books about the supposed nature of the world and its animals and how they should behave. Less often are we reminded of what Aristotle told us 2,300 years ago: 'one should study nature in the individuals behaving according to its rules and not in the corrupted ones'. Once in the forest, I realized that chimpanzees had not read any of these books and simply behave in the way that is the most optimal for survival in the world in which they live. This is a healthy intellectual experience, even if it contradicts many of the ideas proposed in the books. It is thus most important that field workers be faithful to the animals they study and accept that it is more important to describe exactly what they are fortunate enough to observe and not confine their observations to proving prevailing theories or current orthodoxy. Only from reliable and detailed data will our knowledge evolve. Theories are just tools to help us progress, while only observations will teach us what nature is.

When I started, with my wife Hedwige, to study the chimpanzees of the Taï forest, projects studying the Gombe and Mahale chimpanzees in the open woodland/savannah of Tanzania were already in their second decade. We were the newcomers and had to be particularly careful when we made observations that did not fit with what was known from these famous studies. At the same time, the main reason for our project was to study *forest* chimpanzees, given that the vast majority of chimpanzees live, and had always lived, in the depths of the African rainforests. While studying the nut-cracking behaviour of the Taï chimpanzees, we quickly noticed how different they were from what we knew of the Tanzanian chimpanzees. This exemplifies how extraordinarily adaptive chimpanzees are and how fortunate we are to have an animal species so close to ourselves to demonstrate how ecological and demographic factors influence the evolution of many of the most sophisticated social behaviour patterns.

When following the chimpanzees in the forest, we were always aware of their great similarity to us, yet at the same time how very different they were. Chimpanzees are not humans and humans are not chimpanzees. But clearly, with our long common evolutionary history, part of the human is alive in the chimpanzee and, conversely, part of the chimpanzee is alive in us. For some, they are too close for comfort! For us, that proximity provides us with our roots.

In this book, I hope to share with you, readers, the intimacy of the life of the chimpanzees in the Taï National Park of Côte d'Ivoire, how I have experienced it by observing them for 30 years. I will present the complexity of the choices females and males have to make to survive and reproduce in such a dense tropical rainforest (Chapters 2 and 3). Hoping to have raised by then the reader's interest in the Taï forest, I will detail one day in the life of a chimpanzee from dawn to dusk to give you a flavour of what it means to observe animals in their own world and how challenging and diverse life is in the forest (Chapter 4). We will see how chimpanzees in the African forests deal with the predation pressure from leopards, and follow them as they forage through the thickness of vines, cross rivers and swamps to obtain their varied diet, including many different fruits, hard nuts, hidden insects, and mushrooms, and how they hunt for meat, which is natural for wild chimpanzees. The descriptions culminate in the warfare observed in the Taï chimpanzee populations that mirrors some of the most violent human behaviour, while at the same time it reveals some of the most altruistic and heroic facets of the chimpanzee nature (Chapter 5). Throughout the first five chapters, there will be comparisons with what is known from other chimpanzee populations and we will find analogies with humans, as unavoidable similarities or differences will be emerging. However, following my conviction that chimpanzees are mainly tropical rainforest dwellers, I will limit these comparisons so as not to divert us from a complete impression of the life in the forest. Then in Chapter 6, I will proceed to a much closer comparison with other chimpanzee populations, including the famous Gombe chimpanzees observed by Jane Goodall and colleagues, and detail how differences in leopard predation and food abundance produced a wide array of differences in the behaviour of the females and the males, as well as in the prevalence of some often human-like labelled behaviour like cooperation and altruism (Chapter 6). Finally, I grant a closer look at human behaviour and propose that during man's expansion into new habitats in the last 100,000 years, competition for sex has actually become more violent and destructive, and that in this process women have lost much of the freedom they previously had (Chapter 7). I hope by the end to have convinced many of you that by exploring our chimpanzee heritage, we can identify some of the factors that made us human. This might then lead to a greater admittance of the negative aspects of human behaviour that we witness daily in our so-called civilized world and perhaps help to mitigate them.

2

Inconspicuous female superiority

In March 1983, two female chimpanzee friends, Malibu and Poupée, were both at the peak of the sexual swellings that mark their sexual interest, and attracted all the males of the North Group. During this period, they led the foraging trips of the whole crowd of eager males. They regularly liked to climb trees just to rest or for no other obvious reason, thus provoking much tension between the males as space to approach the desired ones became limited in the trees. During the whole week when both females were at the height of sexual activity, the conflicts between the two highest-ranking males Brutus and Schubert, the latter already an impressive young challenger of the older males for some time, developed into some highly risky acrobatic charging displays between them, 40 metres high up in the tallest trees from where the females watched the aerial ballet. The older Brutus was clearly at a disadvantage. The wild jumps between the branches were breathtaking and I was impressed that they would risk their lives for some 10 seconds or so of sex. Malibu and Poupée were leading the patiently following males who were busy controlling the movements of the females as well as preventing others from mating with them. It resembled a never-ending cat and mouse play between the males mediated by the females. From time to time, thanks to the dense forest growth with a restricted view to at most 20 metres, one of them would succeed in sneaking away with one of the females and tempt a quickie before being noticed and chased away. At other times, a male managed to approach the females, but then they would scream and immediately the other males would take care of the imprudent. Only Brutus, the alpha, dared and was tolerated a few times to mate with one of them under the watching eyes of the other pretenders.

Was the group of these males able to impose their will on Malibu and Poupée? Was Brutus, the highest ranking one, really able to prevent the others from mating with the two females, who in turn seemed confident enough in their attitude? Were the females really able to choose among those males, who were so much stronger than themselves, either by agreeing to sneak away or, on the contrary, by screaming to elicit a preventive reaction? Brawn or brain? Coercion or choice? Males versus female choice?

PASSIVE FEMALES OR DOMINANT MALES?

Sex was born when gametes became unequal in size. Living plants and animals did reproduce before the advent of sex but did so either by individuals producing offspring without fertilization, a process called parthenogenesis, or by equal individuals exchanging genetic material, so making new ones. By convention, it only became 'sex' when the gametes became uneven in size; individuals producing larger gametes were called females and those producing the smaller gametes were called males. Because in nature an increasing degree of size difference in gametes appears and we know that we started out with same-sized gametes, biologists have viewed this process as the males 'exploiting' females by investing ever less in gamete production while forcing females to invest more. This led to females in all sexually reproducing animals producing large gametes containing a lot of material alongside the genetic information but only in limited numbers, while males produce huge numbers of tiny gametes consisting almost solely of genetic information.

As a result of these differences in energetic investment and morphology, it is expected that male and female behaviour would differ in important ways. The cost of a gamete is very small for a male and, therefore, males are expected to try to mate with as many sexual partners as possible to increase the likelihood of reproducing. On the other hand, an unsuccessful gamete is a larger loss for a female and, therefore, females are expected to be much more selective than males and mate only with 'good' partners, so as to make the best of her investment. This trend should be stronger in mammals as the female carries the foetus in her body and for a long time she is the only provider for the offspring. So, females risk losing more from investing in the offspring of a 'bad' sire.

Thus, sex viewed as the exploitation of females by males was the classical way of looking at the relationship between the sexes. Despite the fact that Darwin had written about the importance of female choice, females have long been considered as the passive sex. However, with internal fertilization, female mammals find themselves in the position of choosing which sperm can fertilize their egg. This key fact has long been underestimated, for the displays of ornaments and weapons seen in males of so many species have so impressed scientists that they have had difficulties in seeing the wood for the trees.

True, male competition is impressive enough. One is left wondering how much scope is left for female choice, when one sees the remarkable fights between males of well-armed species, such as red deer with their gigantic antlers, elephant seals with their imposing sabre-like canines, or gorillas with their massive muscular bodies and heads. So even when female choice was recognized as a potential important force, one doubts that female choice could ever impose itself over the awe-inspiring power displays of males. It was suggested that females would choose the winners of such fights as obviously they had to be the best and so preferred by females. The female choice should be the winner of the male-to-male competition and could not be independent of it.

BUT FEMALE CHOICE IS A MAJOR FORCE!

Alison Jolly (1999) in her book *Lucy's Legacy: Sex and Intelligence in Human Evolution* astutely reminded us that most biologists are men and as such they are too happy to endorse an attitude accentuating males' control over sexuality. Genetic results opportunely came into play to dramatically support her point, as the myth of female passivity began to fade away when observers of monogamous bird species started to realize that the reality was much more complicated than previously assumed. Many birds reproduce monogamously, as the effort required to brood the eggs and feed the chicks is too demanding for a single individual, and a pair is noticeably more successful at raising the chicks. Therefore, monogamy was thought to be of interest to both sexes, and the near absence of sightings of extra-pair mating in these species seemed to substantiate this interpretation. This view was, however, missing the key data about paternity. Once genetic techniques became available, they were first applied to birds, as it was possible to sample small amounts of blood from potential parents and hatchlings. Knowing the genotype of the chick and its mother, it is possible to compare the genotype of the putative father or any other males present in the population to determine the real father.[1]

A shock awaited the macho view, as extra-pair paternities were surprisingly high and, depending on the species, reached 30 to 70% in well-known monogamous bird species. Furthermore, some studies sampling a whole population of monogamous pairs living in adjacent territories revealed that if some males were very successful by having no extra-pair paternity in their nest while fathering many extra-pair chicks in neighbouring territories, other males were totally unsuccessful and had no chicks of their own. Often in these species, males were not aware of the situation and worked hard to feed the chicks in their nests! Females were subtle enough to achieve their goal without force and by doing so in such a way that males, and scientists, were not aware of their behaviour.

It seems that females were choosing their male partners based on two different criteria. First, they looked for a territory holder as there was no way to raise chicks without one. Once they had one, however, they looked for a 'good male' and if their territory holder was not good enough, they would look around for a better one. Females mated with the extra-pair males in such a way that observers did not notice anything at all. Such stories repeated themselves with some nuances depending on the specifics of the life history of the species considered; not all species were territorial, and the rate of extra-pair paternity varied. It is puzzling that only in some bird species, the males seem able to detect extra-paternity and the more doubt they have, the less they will work to feed the chicks.

Large strong males, contrary to small-sized birds, should surely be able to avoid such surprises and impose themselves on females. No! Even in strongly dimorphic species with large harems, the story turned out to be strikingly similar. For example, in the Antarctic fur seal, the huge males establish territories on beaches where the smaller females gather in large numbers to reproduce, and aggressively chase all competitor males away. During the months of the breeding season, the territory holders, although dispersed throughout the colony, remain essentially static, while the females tend to be more mobile, although movements are difficult across the crowded colony and costly because of the high level of aggression towards any territory intruders. Nevertheless, the genetic method revealed that about 75% of the females conceived to males that were not their territory holder. Females were selecting males according to a subtle balance between high genetic diversity and low genetic relatedness.[2] Thus, even in classic examples of harem species with much larger males, female choice can strongly interfere with male–male competition. The sex strategies can be much more complex and subtle than commonly recognized and the image of the females as the passive sex exploited by larger and more powerful males is increasingly inaccurate.

Some basic physiological properties of internal fertilization have paved the way to more female control over reproduction. Both the selection of a male before mating, including the mating with different males,[3] and selection of sperm after mating but before fertilization allow females to make a decisive choice about the sire of which only the first selection stage could partly be prevented by coercion from the males. For a long time, it was thought impossible for females to select sperm, and the term 'sperm competition' was coined to describe the competition of different males' sperm within the female genital tract, where again the female was thought to be passive. The fact that so many different types of sperm shapes and lengths are observed was proposed to reflect this competition between the males in species where females mate with different males. However, at the same time, females could be seen as selecting between sperms within their body so as to select the best sperms. Science is still collecting data to be able to answer these alternatives but some results hint that both might be true. We are still unclear about the precise mechanism by which females could choose between sperm of different males,[4] but the first indication that a female could actually influence which sperm could fertilize her egg was the observation that certain females were observed to produce many more infants of one sex. Thanks to long-term field studies, we discovered that high-ranking females in red deer and some monkey species produced more males than low-ranking females, who tended to have more females.[5] This has initiated scores of studies trying to discover the different ways in which females could impose their preference on the males.

WHAT ABOUT FEMALE CHOICE IN CHIMPANZEES?

Jane Goodall (1986) suggested that Gombe male chimpanzees are powerful enough to coerce females to mate and she felt females had no way to express their choice. Recent observations with chimpanzees living in the Kibale forest of Uganda[6] tend to support the same kind of conclusion, clearly suggesting that male coercion, as seen in many other mammal species, is an important element for understanding the distribution of mating in this population.[7] At the same time, Alison Jolly (1999) in her book *Lucy's Legacy: Sex and Intelligence in Human Evolution* and Sarah Hrdy (1981) in her book *The Woman That Never Evolved* remind us not to always accept things at face value but to look more carefully. The main point is that females might be wise not to oppose sheer force, as this might expose them to physical retaliation from the males. This has been coined 'cryptic female choice' by some, and there the devil is in the details. Were Poupée and Malibu coerced into accepting the dominant males, or were they able to choose? What importance has female choice in Taï chimpanzees?

With their large testes, male chimpanzees are almost constantly ready to mate with a sexually active female. In addition, the large pink sexual swelling of the females, that is the skin of her genitals that swells conspicuously near ovulation, attracts all males of all ages in the group. It is not uncommon to see a fully swollen female accompanied by a large group of males, all escorting her into the forest. Then she is the one indicating travel directions and timing of foraging. Males seem to have totally forgotten their independence and, for an extended period of time, the competition between them is aimed at mating and preventing matings. The principal game consists of attracting the desired one out of sight of the group and to copulate rapidly before any one notices the trick. At other times, there is greater tolerance between the males and many are able to mate without any overt aggression. Thus, chimpanzees seem to be quite promiscuous, with females mating with several males, and it seems difficult to see what kind of choice they really have. However, we should not forget that this is of extreme importance to females and with internal fertilization they have the last word.

Yet, it seems the preferences of the Taï females for some males directly influence male mating success. The genetic study of paternity in Taï chimpanzees revealed a surprising 10% of extragroup paternity, suggesting female choice. All male chimpanzees are very aggressive towards stranger males and we never saw individual males visiting the study communities as we sometimes observed with females. So, it seems likely that extragroup paternity could be a reflection of the female's active role in selecting sires. To better understand this specific aspect of the female sexuality, we started a detailed study of sexual behaviour in Taï chimpanzees. This showed female preference for certain males to be successful as they were able to prevent disliked males from mating with them.[8] In a quarter of all sexual encounters, the females proposed to the males and thereby revealed strong preferences for certain males. Interestingly, these preferences were only marked during the fertile phase of their sexual swelling period, that is, near the day of ovulation, while at other times, still fully swollen but less fertile, they were much less discriminatory. Amazingly, the males seemed to perceive the female preference and when they proposed mating to females, they did so in agreement with this preference. Males' sexual initiatives mirror females' preferences. When they were not in agreement with their preferences, the females resisted the male attempts and were very successful at blocking them. Males basically accepted female choice and did not coerce them into mating against their will.

Female control of male sexual success was accentuated furthermore by the fact that 93% of mating was interrupted by the females. The duration of copulation was about a third longer when females mated with their preferred males than when they mated with less preferred ones. It was not uncommon to see a female darting away from a male after a few seconds during a mating before or during ejaculation. If ejaculation does not take place, the male may reinitiate copulation, and the same scene may happen three or four times.

Being selective when fertile makes a lot of sense, whereby females ensure that their preferred male is likely to be the father of their infant. But why are females less selective when less fertile? One reason is that the lifetime of the sperm is pretty short, about three to a maximum of four days in humans. In other words, the sperm does not survive for long in the female genital tract and her risk of being impregnated by a male that mated with her five or six days before ovulation is particularly low. Knowing that infanticide by males has been observed in chimpanzees, this tremendous cost to a mother should be avoided by all means.[9] The more males have mated with her, the less certain any one of them can be of not being the father and the more tolerant they will be with her offspring. This will directly contribute to increase the chances of survival for her infant. She has to achieve a fine balance between being confident of being sired by the preferred male and making as many as possible others 'believe' they might have fathered the infant. Female chimpanzees by being less choosy in the early days of their swelling phase and more choosy later, when ovulation is nearing, seem to achieve just that. This mixed strategy of the female gives her the best solution in a complex situation.

Females may go out of their way to find the right sires. If they feel that there is no 'good' male within the group, they could visit neighbouring groups searching for one. One evening on 6 June 2002, a group of nine adult chimpanzees of the South Group eating fruits in a fig tree were joined by Nadesh, an adult female from the Middle Group with a fully swollen sexual swelling. Nadesh was a central female of the Middle Group and had two offspring when we first met her in 1997; her daughter was already 8 years old and her son 2 years old. Genetic analysis revealed that the oldest male of this small group was the father of her son. In the few years following 1997, the Middle Group suffered a rapid decline in the number of males; two disappeared without obvious reasons, while the last one died from anthrax. Sadly, Nadesh's son died during the same anthrax outbreak on 13 February 2002. Following these mortalities, the Middle Group had only one male left, and probably as a consequence Nadesh started to visit neighbouring chimpanzee communities. That evening of 6 June, Nadesh mated with five males of the South Group and stayed the night with them. She stayed another day and a half with them despite the growing aggression of the resident females. She came back a week later for a day and a half mating mainly with the dominant male. She then again became a stable member of the Middle Group. Ten months later, Nadesh gave birth to a male baby that survived for only 17 months and sadly disappeared before we could get some samples from him. Five months after the death of her baby son, Nadesh was again seen visiting the South Group for nine consecutive days in May and July 2005 and mated 15 times with the males of that group. The aggression of the resident females continued all the time and the support that Nadesh received from the resident males decreased over time so that it appeared that on a few occasions she left the group to avoid these females. Nevertheless she stayed for nine days, showing that she was determined to get a sire.

Intriguingly, in October 2005 Nadesh was absent from the Middle Group and not observed visiting the South Group. So, it seemed she was visiting another community, possibly the one in the east, as we knew that this was a large group. Nadesh seemed to have precise knowledge about her neighbours.

In January 2006, Nadesh gave birth to her youngest daughter, Naplo, exactly 8 months after her first visit to the South Group, which corresponds exactly to what we know of gestation time in chimpanzees!

Visits to neighbouring groups by mature females have been seen consistently in Taï chimpanzees. This is not easy to observe as normally females non-habituated to human observers will immediately run away when they see them. In each of the three study communities, we saw females, sometimes with babies, visiting for some minutes up to some hours. We cannot judge how long they would have stayed if we had not been around, but such visits have never been reported for any other chimpanzee populations.

FEMALES' MANIPULATION OF MALES

As mentioned above, when Nadesh appeared in the South Group she had a full sexual swelling which directly attracted the interest of the males of that group and they immediately supported her against the aggression of the resident females. If the sexual swelling is a signal for sexual activity that attracts males, it would be useful to females to show it when in need of support, regardless of whether they are ovulating or not. In other words, the sexual swelling could acquire a function as a 'social passport' that females could use in difficult periods of their life. In all chimpanzee populations, females leave their natal community before adulthood and migrate into a new one. They do so only when they have a fully swollen sexual swelling that immediately grants them the support of the males of the new community. For the first months after entry into the new group, they will be almost permanently swollen, but this decreases as they become better integrated into the group. One new immigrant female, who I named Imi, migrated when still quite small; I guessed she was just 10 years old. After being in the group for 11 months and better integrated, she stopped having sexual swellings for 24 months. It was only once she was grown that she restarted her reproductive cycle and presented sexual swellings again. It is well known that young female chimpanzees have a period of adolescent sterility during which they are sexually active but before they have started to ovulate and, therefore, are not fertile. However, Imi was an extreme case of this. Are female chimpanzees really able to manipulate the duration of their sexual swellings to influence the behaviour of the males?

As mentioned earlier, in a large group of females, there is a lot of competition between them and the low-ranking females suffer the most. Intriguingly, young low-ranking females have sexual swellings almost all the time. Fossey, a low-ranking female of the North Group, successfully gave birth on 12 November 1993 to her second infant, a sweet girl we called Fédora, and she immediately took good care of her. However, 4 months later she started to present a sexual swelling, and continued to do so each month during the following years while continuing to breastfeed Fédora for more than 3 years. During this time she accepted all the males willing to mate with her. On average, young and low-ranking mothers resumed production of sexual swellings around 7.6 months after giving birth but they remained sterile for at least another 3 years. High-ranking females, on the other hand, resumed their sexual swellings 33 months after giving birth. Interestingly, this difference did not affect the interbirth interval and the next infants were born at about the same time in both types of females. So, when necessary, females seem able to manipulate their sexual swelling. Fully swollen sexual swellings attract the interest of the males who were seen to mate regularly with all of them. Young males, particularly, will guard and mate frequently with non-fertile females with sexual swellings, and in any conflicts, these males will support the females. Therefore, by manipulating their sexual swellings, females fool males by stimulating their interest, which helps them to cope better with the intense female competition that exists in the Taï chimpanzee communities.

Sexual swelling in chimpanzees can be viewed as a social passport that signals both readiness to mate and high fertility. Hormone studies in Taï chimpanzees have shown that this latter aspect is deceptive, as ovulation takes place on very different days within the swelling period. Predicting the most fertile days within a cycle is, therefore, unreliable. In most mammals, the fertile period of females is very short, sometimes even restricted to just a few hours, as in the marmot or the bison. Thus, we see in mammals a trend towards longer periods of sexual activity during the whole of one cycle that culminates in female primates with chimpanzees that have swellings lasting about 12 days. This trend makes prediction of ovulation more and more difficult for males and allows females to allocate their mating in such a way as to balance their need for assuring a good partner as a sire with at the same time making many males able to mate with them. So the manipulative skills we see in female chimpanzees are not new and culminate in humans where ovulation is totally concealed and female sexual activity uninterrupted.[10]

Female chimpanzees can test males in yet another way – penis length. If the large pink sexual swelling is conspicuous, at the same time it makes it more difficult for a short penis to reach the fertile zone within the female genital tract. The larger the swelling, the less effective is a short penis. On average, the vagina is 5 centimetres deeper with a full swelling and some males have reduced fertility in such situations. In addition, variations in swelling size are impressive, as some females have swellings three times larger than others[11] and, therefore, successful copulation with such females can only occur if the females cooperate by favouring the positioning of the males. It has often been thought that penis length evolved as a way to displace the sperm of other males, but in this case, it is clear that males need to place their ejaculates optimally to ensure adequate sperm transport.

Thus, we observe a mix of morphological and behavioural traits that makes female chimpanzees better able to control their sexuality. On one side, the duration and the size of sexual swellings make the detection of ovulation more difficult for males in all chimpanzee populations. This makes female chimpanzees, in general, potentially able to select their sire more easily than in species with very short sexual activity periods. In addition, Taï chimpanzee females seem able to bring to fruition these potentials so that male initiations mirror female preferences. I am not suggesting that the female does any of these intentionally or in full knowledge of the consequences, but these differences in female control have evolved and they allow females to have better control over the paternity of their young.

THE POSITION OF THE FEMALE IN THE CHIMPANZEE SOCIETY

This central role of females in selecting the possible sires of their offspring, despite a large investment of the powerful males in controlling this aspect to their advantage, is intriguing as it goes against the wisdom of assumed female passivity. To understand how it was possible for Taï female chimpanzees to reach such an important role in the sexual domain, I propose to have a close look at the society in which they live. Sexual interactions cannot be disconnected from the social reality, as the position the females gain in the sexual domain might be directly resulting from their position in the society in general. In other words, do we see aspects of the female sociality in Taï chimpanzees that can explain why they have such an influence on male sexual behaviour? To understand this, we have to become familiarized with more of the intimate details of the social life of the chimpanzees.

Chimpanzees are about as dimorphic as humans, with the females on average 20% smaller than the males. Male chimpanzees invest a lot of time and effort in maintaining a very clear-cut dominance hierarchy, in which all the males dominate all the females and each male has a determined position. The highest ranking male, the alpha male, is able to displace all other individuals from a food patch and is greeted regularly by all group members. Within a chimpanzee group, males are very active, leading group movements, displaying loudly by drumming on tree buttresses that resound through the forest and by charging through the undergrowth intimidating fearful individuals. If the vast majority of these displays are bluff charges not leading to any injuries, they nevertheless interrupt foraging females who need to be constantly alert so as to move out of the way of charging males. If they are not fast enough, they risk being chased for hundreds of metres and hit and slapped if caught by the male. In the worst cases, they are injured, with large cuts in the sexual swellings, and fingers or toes bitten off.

In all chimpanzee populations we know the majority of the females leave their natal group, before reaching maturity, for another group where they will reproduce. Then female chimpanzees are either nursing a baby for up to 4–5 years or are sexually active. This places a special burden on the female, as she has to care for dependent offspring for over 80% of her life. In addition, weaned 5-year-olds still remain constantly with their mother and she waits for them, supports them in conflicts with peers and shares food with them. When the infants are about 10 years old, they enter adolescence and start to leave their mothers, but they still spend about 75% of their time in association with their mother. Therefore, a typical view in the forest is that of an adult female foraging with her two or three offspring. Not surprisingly, she is too busy with her progeny and has no time to fight to settle every kind of squabble. A situation very different from that of the males.

Females, however, will fight for food. In Taï chimpanzees, females fight over food once every 5 hours and these conflicts represent over half of all their conflicts. Having to breastfeed a baby and share food with juveniles, it is not surprising that they should fight when it comes to valuable food resources. Meat is one of these and some females are so keen to obtain a share of meat that they gain precedence over most of the males in the group and succeed in securing large portions of meat. Brutus, the long time leader of the North Group, was the best hunter in the group and a generous meat provider for all group members. He had a supporting network of females. Ondine, Salomé and Loukoum always obtained meat from Brutus and at the same time helped him to secure the prey whenever he was implicated in a conflict with other males during the long meat-eating sessions. In addition, in a group of nine adult males, Ondine and Salomé were able to assert their position next to the two highest ranking males against all the other males.

Taï forest is a rich tropical rainforest in Côte d'Ivoire that is inhabited by large numbers of ten species of primates and large mammals, such as forest antelopes, bushpigs and elephants. Not surprisingly, leopards are also present in large numbers. Leopards are big cats with surprisingly powerful jaws. One day, I heard a very loud and sharp scream from a chimpanzee followed by a juvenile's distress calls. I walked carefully in the direction of the calls to find Salomé, the second-ranking female of the group, lying on the ground and her son, Sartre, some 10 metres up in a tree, alarm calling while looking both at her and around about him. I got closer and realized that Salomé's chest had been broken opened by a single bite of a leopard. With one bite, the leopard had broken her ribs and punctured her lungs, so that she died immediately from a pneumothorax. Salomé was a fully grown female, some 150 centimetres tall and weighing 38 kilograms. I waited half hidden near her body and 45 minutes later a female leopard approached her, but vanished on seeing me. Female leopards are about half the size of the males, but she still killed an adult chimpanzee with a single bite! Sadly Salomé was not the only victim of leopards during the years we followed the chimpanzees in the forest.[12] Another day, Héra, a female with a 2-year-old baby and a juvenile, came back into the group after 4 days away with fresh cuts all over her face and shoulders. She was otherwise healthy but her baby was missing. From the cuts, it was clear that she had been attacked by a leopard and she had not been able to save her baby.

This constant pressure from leopard attack is an important reason for chimpanzees to seek out other chimpanzees, so they might receive help whenever it is needed. One day I was in the forest with Ondine, Salomé and two other females all carrying infants on their bellies, when we heard a short call some 100 metres ahead. On hearing the thunder-like call, Ondine rushed in that direction, followed by the others, all running and making loud and aggressive waa-barks. Immediately, the forest around us resounded with aggressive chimpanzee calls, and all group members within hearing distance reacted to the first call. Within seconds, they had joined Ella, a female with two sons, who was in a tree and covered with fresh cuts over her face, shoulders and rump. Just in time I saw Brutus, Macho and four other males, all with their hair erect, rush under her in the direction that the leopard might have taken. For several minutes, the whole forest resounded with aggressive barks as the chimpanzees presumably chased away the leopard. There was an area on the ground where Ella's fight with the leopard had taken place. The outburst of calls and barks, however, was a clear signal to the leopard that it was time to escape, thereby sparing Ella from more serious injuries. Ella could barely walk afterwards and needed 3 weeks to recover fully. Had Héra been close enough to a group of chimpanzees, her baby might have survived.

FEMALE FRIENDSHIPS

The lesson is clear, you fare better within a group than alone.[13] Taï chimpanzees are associated with more group members than other known chimpanzee populations and this is especially true for females. In addition, they remain more in auditory contact with one another, allowing for faster support in case of urgency. None of the other habitats where chimpanzees have been studied has as many leopards as in the Taï forest and most have none. So this one ecological difference is the reason for a suite of differences in the behaviour of the chimpanzees. Taï females are more social and because of larger groups are more competitive with one another for food. One way to alleviate this pressure is by having friends;[14] that is, friends in the sense of group members that support you in aggressive situations and share food with you. All the high-ranking females in Taï chimpanzees had friends. At a time when there were more than 25 adult females in the group, 25 to 50% of the females had a close female friend and certainly all the highest ranking females had female friends. Ondine and Salomé, the two highest rank-ing females in the North Group, were seen together most of the time and when they lost sight of each other they were seen to actively search for their friend and whimpered when not successful. These female friend-ships lasted for years and were extremely stable. Loukoum, another young adult female, had Gauloise as a very strong and stable friend. They were seen together 80% of the time during the 4 years we were able to observe them. In one dramatic instance, Loukoum and Gauloise challenged Ondine and Salomé for more than 20 minutes, and the dominant males tried to calm the situation by charging all of them. But the coalitions were too strong, and after some fruitless attacks the males simply screamed without doing anything, while Ondine and Salomé were pursued by the younger Loukoum and Gauloise. Such friendships were only interrupted by the death of one of the partners. After Gauloise's death, Loukoum decreased progressively in rank and was never seen to have such a close friend again. Ondine survived her friend Salomé's death, and within minutes she had adopted her orphaned son Sartre, which illustrates an additional benefit of friendship. Following the death of Salomé, Ondine retained her position as the alpha female for some time but lost it eventually to Poupée, who had the support of her friend Malibu.

Friendship provides stable coalition partnerships which puts females in a better position during social conflicts. It is important to realize that females also compete with males for food. I saw many coalitions where females supported males, as in the case of Ondine and Salomé supporting Brutus, where he clearly counted on them in many instances. This was very clear during the intense competition for meat, when Brutus and Ondine acted as a team and whenever Ondine was forced to scream, Salomé would come to support her. While Brutus was charging one of the males, Ondine and Salomé would remain with the dead prey and protect it from any pretenders until Brutus came back to reclaim ownership and control the meat sharing. This was an important support for Brutus and was seen to extend to other social situations besides the meat conflicts. Friends fought much less with one another than with other individuals. In addition, friends were seen to intervene to help their partner in restoring peace after the latter had had a fight with an opponent. Such interventions by a friend would be accepted by the opponent and the relationship between the two hotheads would be re-established at the same level as before the fight.[15]

The more you need them, then the stronger the relationship with your friends. In larger groups, the association of female friends was very high, while the association decreased by about a half in smaller groups. However, the association between them remained extremely stable lasting for up to 7 years and interrupted in 86% of cases only by the death of one of the female partners. In the Taï forest, the high sociality of the females led them to acquire stable friends for the whole duration of our study and this was only marginally affected by the size of the community.

When no leopards are around, as in the Gombe National Park in Tanzania or in the Kibale National Park of Uganda, female chimpanzees are much less social and spend most of their time alone foraging for food with their offspring. Males are more aggressive with the females and are seen to attack them regularly, slapping and hitting them for no obvious reason, even when they are not sexually active. It is proposed that males impose their will on the females that are not in a position to implement their preferred choices.

MOTHERS' FIGHTS FOR THEIR SONS

The female's investment in reproduction does not end with the choice of sperm, and, in chimpanzees, mothers invest for much longer than the duration of milk production. The females, after selection of a good sire, need to ensure that their choice will be successful. In Taï chimpanzees, high-ranking mothers were seen to have a new baby 2 years later if they have a son than if they have a daughter, while low-ranking females invested 11 months more in daughters than in sons.[16] A result of this longer investment and later weaning for the sons of dominant mothers was that these sons had better survival rates to adulthood than all other offspring. The most likely explanation for such an investment is that sons of high-ranking mothers have a better chance of becoming high-ranking males and as such will have more offspring.

Thus, the greater investment of the mothers makes sense. But surviving to adulthood is only half of the story.

Ella was one of the shyest females of the North Group, and I saw her only irregularly and initially gave her a peripheral status. In fact, Ella was a dominant female and it was our presence that intimidated her. It took many years and extremely cautious approaches before she lost her shyness. She was always with her two sons, Fitz and Gérald, born in 1975 and 1983 respectively. Family ties were strong. Fitz showed no desire to leave his mother until he was 13 years old. As he grew older, we saw her progressively more regularly in the group of the adult males. Then we realized that Kendo, born around 1969, must also be her son; he was surprisingly tolerant towards her, letting her take large pieces of meat and fruit from his hands, and he was very relaxed with her two youngsters. Kendo too had a strong physical appearance and was certainly the fastest walker in the community. He showed a keen interest in climbing the social ladder. He often displayed violently towards the adult females, regularly challenging the most dominant ones. We even saw him twice challenge Brutus, still the alpha male, biting off a bit more than he could chew.

In late 1986, it first became obvious that Ella was interested in helping Kendo to improve his social status. He was trying to dominate old Falstaff and each time that Kendo screamed in front of Falstaff Ella rushed to support Kendo, followed by her two younger sons, all three of them barking. The sight of this rather impressive trio must have had its effect, as Kendo rapidly dominated Falstaff. Beginning in 1987, Ella started to be a regular member of the males' group, and Kendo was by that time third in the male dominance rank order. He was dominated only by the ageing alpha male Brutus and Macho. Fitz was clearly very interested in the males' social relationship and the 5-year-old Gérald was also a keen watcher of these interactions. In October 1987, the number of conflicts between Macho and the Kendo–Ella team increased markedly. Three times the conflicts were clearly initiated by Ella herself. Kendo rushed to rescue his mother, and together they chased Macho away. It became apparent that Macho felt uneasy with Ella, uncertain about how to handle this female who was always accompanied by Fitz and Gérald, making a strong team, and always obtaining the support of Kendo whenever she cried long enough. Now, for the first time, Kendo was able to retain an adult red colobus prey in front of Macho when Ella was present. In early summer 1988, Macho became the alpha male of the community. Brutus gave up his position without much resistance, at about 36 years of age. Now

Macho used the strategy that he himself had suffered from Brutus some years ago: he fiercely challenged Kendo whenever his mother was absent. Kendo clearly did not feel strong enough to challenge Macho, and we gained the impression that he was uncertain about his mother's ambitions for him. He sharply increased his association with her during the summer of 1988, and both of them kept away from the group.

Beginning in 1989, Kendo tried to disrupt the bond between Macho and Ulysse by associating with Ulysse, but he did not have much success – Macho pressed Kendo very hard and slapped him regularly. During the summer, Ella's whole family was absent from the group as Ella gave birth to a fourth son, Louis, in August. In September 1989, the family reappeared in the group. In November, despite her tiny baby, Ella started to support Kendo actively in his challenges against Macho. This time, Macho clearly feared Ella's support, as we heard him scream whenever he heard her supportive barks, even if she was out of sight. In December, Kendo defeated Macho in a violent fight in which Macho lost a toe and the phalanx of a finger. Kendo became one of the most powerful and aggressive alpha males in the community that we had ever seen. His mother died during the summer of 1990, but she had achieved her goal better than she could have imagined as Fitz gained dominance over Kendo without a fight. In the 4 years when both brothers were dominant, they sired seven offspring, Kendo fathering five of them. Clearly, Ella's strategy of investing heavily in her oldest adult son had been successful.

The example of Ella is just one, although the best documented one, from Taï female chimpanzees, but similar maternal investments in the future development of adult sons has also been described in the chimpanzees of Gombe with the famous Flo that supported her son Figan to become the alpha male of the community and later Fifi doing the same for her own adult son Frodo. Similar observations have also been reported from the Mahale chimpanzees. To be supportive, the mothers have to survive until their sons are old enough to challenge high-ranking males. Therefore, such observations are not abundant, but are seen in all chimpanzee populations.

THE PUZZLE OF FEMALE POWER

A female social network is obvious in most primate species, where females remain in their natal group for their whole life. Matrilines are formed whereby daughters gain social rank directly adjacent to their mothers. Females within such matrilines support one another

and coalitions are regularly observed. Such female-based social networks remain stable for generations, and the dispersing males enter
and leave such groups without affecting the social organization.
Female philopatry is the rule in a majority of primate species, such as
macaques and baboons, and directly influences the social relationships
within the group. This does not, however, imply that females dominate
males, and males often through sheer strength fight their way to the
top and keep control of a female group for a number of years before
being replaced by stronger and younger males. An extreme case is seen
in the langurs of India where female networks remain stable for years.
One male controls the female group and gains exclusive access to all
females. Every 2 to 3 years, the male is expelled by a younger male who
will immediately kill all the young sucklings of the females in the
group. In such a way, the male during his relatively short tenure of
the group will protect only those infants that are his own. Despite the
fact that there are many more females, they do not seem to be able to
prevent the male from killing their babies.[17]

Males are expected to dominate females even more in species
with male philopatry as not only do they have the advantage of larger
size but, in addition, they have the advantage of long-term social relationships. This is what is expected in chimpanzees, humans and other
male philopatric species. Therefore, our observation of strong female
social relationships and control over much of the sexual relationship is
unexpected. In my view, that is where the long life cycle of chimpanzees
comes into play. If it is true that females migrate, once they have
migrated they stay in their new group for decades and, therefore, have
plenty of time to develop strong and intimate relationships with other
non-kin females if living conditions make them profitable. Thus, it is
the interaction between a long life and high predation pressure that
gives rise to the female position in society, including strong friendships
and more control over reproduction.

In sum, the chimpanzee females' panoply of behaviours to
control reproduction includes swelling size, hidden ovulation and
imprecise fertile periods, selective male preference, high success in
copulation refusal, and a decisive influence on male mating initiations
and mating duration. All these seem to coincide in assuring females'
control over the paternity of their infants and providing an optimal
environment for their survival. This, combined with a sex-specific
longer term investment, allows females to influence the success of
their infants by taking into account their specific social position. All
this happens within a society where the males are systematically,

through sheer strength, dominant over all females. However, when it comes to reproduction, female chimpanzees can assert their interest and impose a choice that might go some way to opposing the males' interests. As we shall see in Chapter 6, the situation of the Taï females is in this respect quite different from other chimpanzee populations and this underscores the role some specific ecological factors can have in the life of individuals.

NOTES

1. The genetic tests for paternity use microsatellite markers that are tandem repeats of bases on the nuclear DNA without any known selective function but subject to a relatively high mutation rate making them very variable. Therefore, they are perfect neutral markers of the individual, a kind of chemical fingerprint. If enough of such variable markers are used, on average 9–12 loci, a certainty of exclusion of well over 99% is achieved. That is, a male can be excluded as a father when there are one or two mismatches between the male and the infant taking into account the genotype of the mother (we know the mother gives half of her genetic material to her infants and so the father has to provide the missing half). The genetic method works only by exclusion. That is, the male considered as father is the single one sampled not excluded and, with a high enough exclusion probability, we can be confident about the result.

   One of the central difficulties in doing paternity testing is to be sure that you have sampled all potential fathers. Missing the mother decreases strongly the power of such genetic testing, but not having samples from all possible fathers makes the test even shakier.

2. The female choice of highly divergent but more compatible males has been supported by more and more evidence in recent years for species as different as fur seals, marsupials, field crickets, savannah sparrows, blue tits, bluethroats and sand lizards, and possibly humans as well (Johnsen *et al.* 2000, Hoffman *et al.* 2007, Edvardsson *et al.* 2007). In a few cases, like the blue tit, it could be demonstrated that the young with a more diverse genotype survived better than those with less diversity. It is thought that this favours the immune system of the offspring and makes them fitter. By choosing highly divergent males but with some relatedness to themselves, females at the same time preserve some of the local adaptive genetic advantage that has been acquired.

3. Theory predicts females to be more selective than males because they have invested relatively more in their eggs than males in their sperm. It therefore came as a surprise to realize that, in many species, females mate with many more males than is necessary to conceive. Multimale mating by females is relatively common among mammals, occurring in at least 133 species, including 33 species of primates (Wolff and Macdonald 2004). Detailed studies showed that females often increase their mating with other males to select males more genetically divergent from themselves and thereby increase the viability of their offspring (e.g. monogamous marmots, elephant seals, pied flycatchers, etc.) (Cohas *et al.* 2006). Furthermore, offspring quality, in terms of survival, has been shown recently to increase in females having more male partners (e.g. antechinus, a small carnivorous marsupial, field crickets, savannah sparrows, blue tits) (Mays and Hill 2004, Edvardsson *et al.* 2007).

4. Scientists were, however, already aware that females could manipulate the number of sperm in their genital tract by expelling some of them. In many species, it was observed that after mating a certain amount of ejaculate could be rejected by the females, for example, in zebras, elephants, chimpanzees, humans and others. Specifically, sperm of low-ranking males is expelled more systematically than is that of dominant males in species like feral jungle fowl (Pizzari and Birkhead 2000). In addition, morphological studies of invertebrate female reproductive tracts showed that some specific morphologies allow for greater female control: for example, sperm storage organs where female can store some sperm for later fertilization, or specific tubes where sperm are placed in such a way that either the first or last placed sperm would always be used for fertilization (Andersson 1996, Eberhart 1996, Dixson 1998).

5. The first study was with a red deer population on Isle of Rum, Scotland, which was studied for many decades and showed that high-ranking females were having many more sons than low-ranking females and that the sons of high-ranking mothers developed more quickly and strongly, thus having a good life start (Clutton-Brock et al. 1984). Such a phenomenon had been predicted by some scientists who were theorizing about the parent investment theory (Trivers and Willard 1973), but confirmation came as a surprise. Since then this has been confirmed for different primate species (Dunbar 1988) as well as some bird species (Komdeur et al. 1997). The exact mechanism of how females can distinguish the sex of sperm or other qualities generally remains unknown, although some have proposed that the surface properties of sperm can be evaluated by the ovule before fertilization. Other mechanisms have also been proposed but empirical confirmation is still awaited (Eberhart 1996).

6. Chimpanzees have been studied since 1960 in Tanzania, in Gombe National Park by Jane Goodall and colleagues, and at almost the same time by Toshisada Nishida and colleagues in the Mahale Mountains National Park some 200 kilometres south of Gombe. Three more recent detailed studies on chimpanzees in Uganda are still under way in Kanyawara by Richard Wrangham and colleagues, in Ngogo by John Mitani and David Watts, both in the Kibale National Park, and the last in the Budongo National Forest by Vernon Reynolds and colleagues. A few other populations have been studied, and will be mentioned later in Chapter 5, but all were of a shorter duration and provided less detailed information on the individuals and their behaviour.

7. Multiple mating by females with different males seems to be promoted by the presence of sexual swellings in different species of primates, such as in different species of baboons, colobus monkeys and chimpanzees, and is often associated with a higher level of aggression in the males (Smuts and Smuts 1992, Clutton-Brock and Parker 1995). This has been proposed to represent male attempts to coerce females in mating with aggressive males and limiting female choice. In the chimpanzees of Kanyawara in the Kibale National Park the sexual interactions were followed and showed that males mated twice as frequently with females towards which they were the most aggressive (Muller et al. 2007).

8. Rebecca Stumpf closely followed the Taï chimpanzee females to see how their behaviour changed during the sexual cycle (Stumpf and Boesch 2005, 2006). By recording who initiated the sexual interactions and how the other responded, it was possible to distinguish between female initiation reflecting female preference, and male initiation reflecting male preference and possible coercion. This gave a precise knowledge of how the sexual interactions developed.

9. Infanticide has been proposed as a driving force to which females have to adapt. For example, in lions and langur monkeys, solitary or groups of males take over groups of females by ousting the resident males and then they kill all infants so that females quickly come into oestrus (Schaller 1972, Hrdy 1977). In this way the new male(s) do not invest in infants from other males and gain more offspring for the same length of tenure. Even in stable social groups, like chimpanzees, males from within the group have been observed to kill infants of mothers that seemed either to have copulated too frequently with low-ranking males or were born during a period of strong competition (Nishida and Kawanaka 1985). Females have been seen to react to this threat by faking receptivity and mating with the new males even if they are already pregnant. Paternity confusion seems to be the main but not always successful weapon females have found to counter more powerful males.

10. Concealment of ovulation has progressively become more important in mammals and seems most complete in humans. Often, anthropologists have explained it in humans as a way to make monogamy attractive to men because of the constant enjoyment of sexual activities. Since this trend is general in mammals, and monogamy is not obvious in most cases, the need to confuse paternity to potentially infanticidal males may also have played a large role in the evolution of human sexuality (see Chapter 7).

11. Tobias Deschner measured the size of the swellings of most females in the South Group and by analysing video pictures could make precise measurements (Deschner *et al.* 2003, 2004). At the same time he measured hormone levels in the faeces of the same females, and gained detailed knowledge of the morphological changes in the swellings in relation to the hormone cycle and ovulation.

12. Predation by leopards on chimpanzees was seen throughout our project. A study of leopards showed that they are attracted by the calls of the chimpanzees and they approach and wait nearby for an opportunity to catch one (Zuberbühler and Jenny 2002, F. Dind unpublished data). The majority of the attacks on adults are not fatal and chimpanzees seem able to protect themselves from the first strike by folding their arms around their head and then escape up a tree with cuts on both hands and around the head and shoulders (Boesch and Boesch-Achermann 2000). Leopards are specialized carnivores and cannot afford to risk an injury, as they might not be able to hunt while recovering and therefore will starve to death, so they will hesitate to continue if the first strike is not successful. The danger represented by leopard attack is constant and chimpanzees are extremely supportive when it comes to leopards and will intervene to chase them away whenever they hear them in the forest (see also Chapter 3, section on altruism). The leopard's liking for chimpanzee meat leads them also to scavenge on individuals killed by disease or during warfare with other chimpanzee communities.

13. Predation pressure is considered to be one of the main forces for group living (Krebs and Davies 1991): the more individuals in a group, the less likely is each of them to become a victim. This is thought to be the main reason for the large associations we observe in fishes, bats and ungulates like wildebeests or primate species like baboons. A second direct effect of being a member of a group is improved vigilance: the more eyes in the group, the earlier predators are spotted and alarm given. This is thought to apply especially for bird associations and small ungulate groups that rely on speed to escape from predators. In addition, active defence against predators is more efficient in groups, as demonstrated in the group mobbing of predators by birds and primates.

14. Friendship in animals is rare and usually develops only between individuals that spend a very long time together. Thus, it occurs only between close kin, like sisters or mother and daughters, or between members of the sex that stay in the same group all their lives, which in chimpanzees means the males. Female friendship in chimpanzees, where the females migrate between communities before reaching maturity, was not expected. This shows how important ecological factors can be in shaping the social system within a species. Female friendship changes with competition, so that in large groups female friends spent 75% of their time together, whereas in smaller groups they were together only 45% of the time (Boesch and Boesch-Achermann 2000, Lehmann and Boesch 2004, 2009).

15. Studies showed that friendship had many special aspects other than simply helping during fights and food sharing. In Taï chimpanzees, friends fought much less with one another than with other group members (Lehmann and Boesch 2009). In addition, following a fight between chimpanzee group members, former opponents 'made peace' by showing affiliate behaviour with one another (de Waal 1989, Wittig and Boesch 2005). In some instances, an individual not involved in the fight might intervene and help opponents come together again in an affiliate way. Only if friends intervened for a partner involved in a fight would the relationship between the opponents be re-established as well as if the partners involved in the fight did it (R. Wittig and C. Boesch unpublished data).

16. Chimpanzee females seem not to be able to select preferably one sex over the other, but higher ranking females have been seen to invest discriminatively more in sons by having the next siblings 2 years later, while low-ranking females tend to invest more in their daughters. In addition, sons of high-ranking mothers survive better than other offspring (Boesch 1997). In Gombe also, high-ranking females have offspring that survive better but no sex-specific investment was seen (Pusey *et al.* 1997). So, similarly to the red deer, high-ranking females invest more in sons that once adult will have more offspring, but they do so later in life as they do not seem able to select between different sperm, contrary to the red deer. Trivers and Willard (1973) predicted that in species where mothers can positively influence the reproductive success of their offspring, the mothers should produce more of them and invest more in them at the cost of the other sex.

17. Sarah Blaffer Hrdy pioneered work with the Hanuman langurs of India and revealed dramatic aspects of the reproductive life of this species (1977). Females try to oppose the killing of their infants and counter it by faking sexual interest as soon as the male appears, but generally, the new male kills most of the babies. Males do not always father all the infants within a group (Launhardt *et al.* 2001), and females, by mating with all males in multimale groups, may enlist more support against stranger males (Borries *et al.* 1999). Such female counter-strategies, however, do not alter the fact that infanticidal males are very successful.

3

The tyranny of the testis

One day in March 1983, when Malibu and Poupée had their sexual swellings and their male suitors crowding at their heels, they started to climb a *Uapaca* tree, and right away the hostility arose between the highest-ranking male Brutus and Schubert, his impressive young challenger. Immediately the smaller and younger Macho joined forces with Schubert. Both young males, with their hair on end, faced Brutus and screamed at the top of their voices. Brutus had to choose between chasing them or remaining close to the females. Falstaff, the oldest male of the group, approached Brutus in what looked like an attempt to quieten him. Brutus stretched his hand towards Falstaff to enlist his support and Schubert now had to face both of them. Falstaff barked at him and Brutus chased Schubert and Macho up the big aerial roots of the *Uapaca* tree. Macho, who was farther away from Brutus, displayed above him. Schubert and Macho returned towards the females, hoohing loudly. Then, as Schubert and Macho threatened Brutus again, he approached Falstaff sitting nearby. As the ill-tempered team approached, Falstaff wanted to leave, but an upright Brutus stretched his hand towards him, looking alternately at Falstaff and Schubert. Undecided about what he should do Schubert hoohed for a short time. Nothing happened. Then Schubert threatened Brutus with an extended arm wave. Falstaff made a reassuring move towards Brutus. Once more, the two older chimpanzees chased the younger team away. Schubert in a wild display rushed through the forest and chased all the chimpanzees up into the trees, while Brutus quickly returned to Malibu and Poupée.

During the week when both females were at the peak of their sexual activities, the conflicts between Brutus and Schubert continued either 40 metres high up in the tallest trees where the females had climbed or on the ground where manoeuvring between the males could develop in its full complexity. In the trees, the older Brutus was clearly at a disadvantage. From my modest human perspective, I shared his concern in risking so much high up in trees. If Brutus was unsure up there in the trees, he regularly succeeded in tricking the younger ones back to the ground where he regained his confidence and could deter their attacks. Their conflicts continued for the whole week; however, neither of them was successful as the girls were impregnated much later. Nevertheless, males seem to be willing to take so much risk to gain precedence over other males. It makes us wonder why males are so risk-zealous. Are they unable to detect when females are receptive? How can social cohesion be preserved with such a high level of competition between males?

Sex is one of the most important driving forces in nature. Some animal species are ready to forgo everything to be able to have sex and reproduce. Salmon undertake a hugely exhausting journey up rivers to reach their breeding grounds in small, clear-water rivers where, after having reproduced for the first and only time, they will die. Similarly, the male praying mantis will sacrifice his body for the female to eat while he mates with her. Thanks to this free meal, she will be able to produce more and fitter offspring. To accommodate this particular way of mating, the nervous system of males is adapted such that centres important for controlling mating have moved out of the head into the abdomen. So, the female can eat the male's head without affecting his mating efficiency!

Mammals living in social groups also face strong sexual competition to secure mates and prevent others from mating with them. Often these males are much larger than the females, as is the case in many mammals including primates. Elephant seal, gelada or hamadryas baboon and gorilla males are about twice the size of the females, thereby gaining a decisive advantage in strength. In addition, males have acquired sex-specific natural weapons, such as long and sharp canines or horns and antlers, which help them in fights with competing males. The result of this competition makes the male the 'risky' sex, risky in the sense that a large proportion of males have very few offspring or never reproduce, while a minority of males produce a disproportionate number of offspring. Harem holders often control a large number of females; for example, the harem size of the red deer is about 12 females, while the average elephant seal harem has 227 females! In mammals, where the sex ratio at birth is always close to 50:50, this means that a large number of males never reproduce. However, this is only the tip of the iceberg.

WHAT THE GENITALS TELL US ABOUT
THE SEX LIFE

In addition to such obvious sexual dimorphic traits, the sexual organs of males reflect the intensity of sexual competition faced by individuals. Group living suffers from overt physical competition and males living in multimale groups would disrupt many of the benefits of social life if their fights became too violent. Hence, such males present fewer sexual dimorphic traits than typical harem holders. However, if access to females is not fought so extremely, it does not mean that competition is not intense during or after mating. Since the males' interest is to sire as many offspring as possible, the level of competition between males remains very high. On the other side, females may want to select the best male and multiple mating may allow this, as it has been amply shown that multiple mating profits females in important ways such as increasing her fertility and the quality and survival of offspring. As a result of both sexes' interests, most of the competition happens during or after mating, either by female cryptic choice, where females select between sperm of different males, or by sperm competition, where the sperm of different males competes in the genital tract of the female.

The greater the competition between sperms of different males, the more advantageous it is to have larger testes. The comparison between chimpanzees and gorillas is illuminating in that perspective. Chimpanzees live in large stable communities with many adult females and males, where females are regularly seen to mate with many males within a day or during their cycle. Chimpanzee males have very large testes and a long thin penis. On the other hand, gorillas live mainly in stable one-male groups with many adult females, and in such harems females mate only with the harem holder. Gorilla males possess much smaller testes and a shorter penis than chimpanzees. If corrected for different body size, the difference is even more striking. In the highly competitive system of the chimpanzees, where males displace one another from the females and females are often seen to mate with different males, mating is often very short, lasting only a few seconds. Males need to produce a large number of sperm that are ready to be inseminated whenever an opportunity appears, and they need to be placed deep into the females' genital tract to have a higher likelihood of fertilizing the eggs. Such long penises also help in removing some of the sperm placed previously by other males.[1] Gorilla males normally do not face competition from other males within the group and can take all the time they need to mate and, because of the small penis, mating requires the complete cooperation of the female to be successful. Within primates, we observe a clear trend for males possessing larger testes and a longer penis when they live in multimale groups than when they live in monogamous pairs or harems. Marmosets, gorillas and gelada baboons clearly have smaller testes than chimpanzees, macaques, mandrills or mangabeys, which are multimale and multifemale group species.[2]

The importance of sexual competition can be directly measured from the size of the testes allowing for body size in social primates. Humans rank closely with chimpanzees by having relatively large testes, much larger than those of the gorillas or orangutans, which live a relatively solitary life. Furthermore, human penis length is comparable to that of the chimpanzee and much longer than that of the gorilla. Thus, both testis and penis size is testimony to our sexual competitive past. Both chimpanzees and humans have faced a long evolutionary history of great sexual competition, much more so than the gorillas and the monogamous gibbons or marmosets.

SEXUAL COMPETITION IN MALE CHIMPANZEES

Are high levels of sexual competition in fact observed in wild chimpanzee populations? Ten years ago, without detailed genetic analyses, such a question was simply impossible to answer. Traditionally, we would have looked at mating frequency and assumed that this reflected paternity success. We recognized the limitations of the method, but it was impossible to be more accurate. Now because of the development of genetic analysis using, for example, samples of faeces, we can answer that question.

Males compete intensively in their group. Competition for food and sex is observed regularly and fights between group members are frequent. We observed about five conflicts for every day we followed the group and, in most cases males are implicated in such conflicts.[3] The large majority of conflicts are over sex and for social reasons and they are mainly provoked by the males. These intense conflicts between males have a cost. The greater the conflict, the more testosterone is produced. Generally, higher ranking males have more testosterone than lower ranking males, as they are the ones making most of the aggressive displays in the group. But for all of them, when attractive females are sexually active, the testosterone level of adult males increases by about 40%.[4] Since the production of testosterone is costly

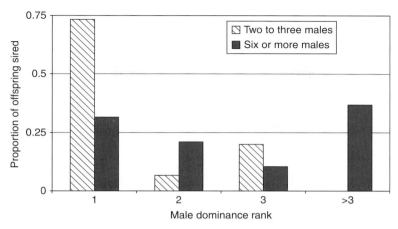

Figure 3.1   The more male competitors are present, the smaller the reproductive success of the dominant male.

to the male as it increases the metabolic rate, decreases fat reserves and has immunosuppressive effects, males suffer a high cost the more they compete for females.

Is paternity also fought for by male chimpanzees? A long-term analysis of the distribution of paternity between the males of three chimpanzee communities in the Taï forest revealed clearly the unequal distribution of paternity among the males: the highest ranking male on average fathers half of all offspring born in the group. However, competition clearly influences paternity distribution, as the more males are present in the group, the fewer paternities will the dominant male be able to secure, while at the same time less-dominant males gain more paternities, as seen in Figure 3.1. This effect is not only due to direct competition between males for the same female, but also because in larger groups more females are sexually active at the same time and, when this is the case, dominant males are not able to control all of them. The dominant male loses over half of paternities when more females are sexually active at the same time, as can be seen in Figure 3.2. Females also have more opportunities to express a preference towards other males.

Therefore, in Taï chimpanzees, a strong effect of male competition is observed and it is clearly stronger in large communities. Competition between males is observed in all sizes of group whereas the effect of the number of females is only seen in large communities, as small communities have only a few active females. At this point, we

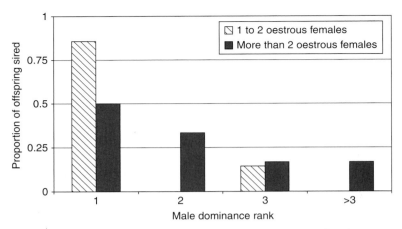

Figure 3.2   The more females are sexually active, the smaller the reproductive success of the dominant male.

could ask ourselves why the highest-ranking males should tolerate the presence of competitors. As we shall see in Chapter 6, male chimpanzees are very aggressive towards males of other communities; therefore, the presence of additional males is beneficial for a community. In addition, and maybe more important, a large number of males attracts females. I saw larger communities attract new immigrant females about once or twice a year, whereas small communities with six or fewer males may wait for many years before any females join the group. So in the end, even if dominant males have a smaller share of the paternities, they still have an absolute larger number of them the larger their communities.

Such a high level of competition makes it clear that each mating is important and the timing of it might be crucial. Female chimpanzees have acquired large sexual swellings that last for 12 days per cycle and are very conspicuous signs of their sexual status. The males are attracted to them but ovulation can happen on any day during a 7-day phase in the second half of the swelling period. This makes the prediction of ovulation, and thereby the most fertile days, very difficult, and provides the female with an opportunity to control the males. Such a period of sexual swelling has been observed in many primate species and generally they last longer than in other mammals, making the prediction of ovulation much less reliable.

However, male chimpanzees seem to have a very detailed knowledge of the physiological changes occurring in the females and realize that the size of the sexual swellings increases slightly towards ovulation. Furthermore, the size of the sexual swellings tends to be larger for conceptive cycles. To complicate matters, each female produces swellings of very different sizes, so the males have to compare relative size and not absolute size in each female if they want to use this indicator to predict ovulation reliably. Taï males seem to be fully aware of this and the alpha male was about five times more likely to associate with a female when her swelling was at its largest, which is during her fertile phase, than otherwise. In addition, he was 11 times more likely to mate with her during her fertile phase. He also increased his association with females as ovulation approached. Therefore, because of his status the alpha male, who has the greatest freedom to mate with females when he wants, adapted his behaviour to the fine-grained signal that a female gives of her fertility and timed his mating to give the highest likelihood of conception.

Males seem to understand a surprising amount about the reproductive physiology of the females. As we just showed, they are aware that not all mating events are equal and that timing is very important. If they really understand that much, do they also have the idea that conception leads to the birth of infants months later? In other words, do males recognize their own infants?[5] The social structure of the chimpanzee community whereby members are almost never all seen together, but only in subgroups of constantly differing composition, makes it difficult for a male to follow all the matings a female may have and reliably infer paternity. Furthermore, sometimes there are as many as five sexually active females at the same time in a group. Despite these complications, Taï males are clearly less aggressive towards mothers carrying their own offspring than mothers carrying unrelated offspring of the same age. At the same time, males play for longer periods of time with their own 2- to 10-year-old youngsters than with unrelated ones of the same age (infants less than 2 years old mainly remain on their mothers and invest little time in play).[6] Thus, Taï males without doubt recognize their own offspring.

Despite fathers' recognition of their own offspring, paternal investment was selective, as they were not seen to associate more with their offspring's mother or support their offspring in fights, but only played with them and showed less aggression towards their mothers. It seems as if male chimpanzees invest both in future reproductions by associating more with females with big pink sexual swellings, as well as in past reproductions by investing in play and grooming with their existing offspring.

COOPERATION AMONGST COMPETITORS

How is social harmony preserved between males with such high levels of competition? The most striking feature of male chimpanzees is that they cooperate so much despite, at the same time, being rivals. This ambivalence between competition and cooperation is a fundamental part of the social life of males. Following the chimpanzees daily provides everyone with the remarkable experience of watching groups of adult males resting in the forest close to one another, sometimes with a shoulder or foot touching, while only a few minutes earlier they had been fighting with wild displays over a sexually active female. This ability of rivals to hunt together and share meat with one another is, in my eyes, the distinctive attribute of male chimpanzee sociality.

Cooperation among competitors requires some specific flexibility.[7] Cooperation is always susceptible to cheaters, as some individuals may want to profit from the rewards of a joint venture without investing in it. In a group of selfish individuals, there is a great temptation to profit from the work of others, and if, in addition, these are your rivals, the temptation might be irresistible. How do you differentiate, for example, between groups of individuals working as a coherent hunting team and groups of parasites who follow a hunter? Both look the same: a group of individuals running after a prey. However, only in the first case are all individuals genuinely trying to contribute to the capture. Detailed observations with groups of lions hunting together revealed examples of the second case where some individuals just followed the hunters.[8] The situation in lions might be easier, since the hunters are females and they do not compete as strongly with one another for reproduction as do the males. For cooperation to work, precise distribution of the benefit is needed to guarantee a reward to contributors during the cooperative act. For example, no meat sharing rules were seen in the Gilgil baboons in Kenya that guaranteed a reward to the hunters. This baboon population was well known for hunting small antelopes. Interestingly, once the dominant male, the main hunter, lost his position in the hierarchy, hunting rapidly stopped as the prey was regularly taken away from him by the new dominant male (Strum 1981).

Meat is an especially concentrated source of nutrients in nature, but obtaining it is very challenging. The 'meal versus life' paradigm explains this, in the sense that prey fight for their lives, while hunters fight for a meal. So, unless you are a specialized carnivore eating only meat, for omnivores there is always an easier alternative to meat. This probably explains why within the primate order hunting is so rare. In addition, primates generally do not have specialized weapons like carnivores to subdue struggling prey.

To acquire the hunting technique is a long, difficult process and so the older experienced hunters are the best. One day, I was following a party of males with two mothers. Brutus was leading and I began to realize that he was looking for monkeys and not for fruits. They had started to move in a steady fashion more or less in a line and did not stop to eat as they usually did. Moving in such a way, they quickly lost contact with other group members and progressed as a tight team. We then heard Diana monkeys calling some way to the north-west. The chimpanzees immediately turned in that direction. In the Taï forest, Diana monkeys are often associated with other less vocal monkey species. As they approached the monkeys, the chimpanzees slowed down and approached silently to surprise them. Not spotting red colobus monkeys, their favourite prey, they searched for a little while under the monkeys looking towards the east and north. Once they were certain that there were no colobus monkeys, they continued their search moving steadily further north. After 10 minutes, they again heard the calls of the Diana monkeys further ahead and, after an exchange of glances, Brutus took the lead. The approach was very careful so as not to be spotted by the monkeys, among which we now could hear the typical copulation calls of red colobus monkeys. Still unnoticed, the hunters started to fan out under the monkeys, the older ones, Brutus and Falstaff, moving forward under the largest part of the colobus group, and the youngest, Snoopy and Darwin, stopping to look up into the trees. Snoopy started to climb some 7 metres, looking both at the monkeys and towards Brutus and Falstaff, who then started to move on the ground further north, while Macho and Kendo moved into positions in between the former two. Macho was upright holding a tree trunk ready to jump once the colobus monkey started to flee. Snoopy, followed by Darwin, continued to climb and immediately the Diana and red colobus monkeys started to give alarm calls and fled towards the north in the opposite direction from where Snoopy was arriving.

The hunt had started and everything speeded up. Macho and Kendo barking loudly rushed up towards the fleeing monkeys. Following the two females, I stayed with Macho whose presence pushed the monkeys towards the north-east. Brutus and Falstaff ran on the ground so as to remain in front of the fleeing monkeys. After 10 minutes of running, Falstaff, with Macho and Snoopy in the trees pushing the monkeys in the same direction, quickly climbed up a tree in front of the first fleeing monkey. His presence in the tree blocked one of their escape routes and the monkeys had either to turn back from where Macho and Snoopy were now coming, or to flee further to the north-west. Some monkeys succeeded in escaping by turning back and running on the thinnest branches 50 metres high up in the tallest tree. Macho and Snoopy lost some time trying to capture one of these monkeys. Most, however, went towards the north-west where Brutus was already hiding halfway up a tree. As so typical for him, Brutus had perfectly anticipated where the colobus would try to flee to avoid the approaching hunters. With perfect timing, when the first colobus arrived in his tree, Brutus rushed up with a hunting bark. Surprised, the monkeys turned back but were now trapped within the triangle made up of Brutus, Falstaff and the Macho/Snoopy partnership. Four monkeys were trapped there and the hunters concentrated on a large mother carrying a baby on her belly. Young Snoopy, now within a metre of her, hesitated, impressed by her threats. Brutus and Falstaff, however, swiftly seized and subdued her, Falstaff leaving with the baby which was already dead. The typical loud capture screams were given simultaneously by Brutus, Macho and Snoopy. The two females who had followed on the ground joined the males in the tree and they all started to feed on the meat. The forest quickly became quiet and Brutus, who controlled the kill, shared meat generously with everyone, and all the other hunters received large pieces. Fifteen minutes later, we heard some more chimpanzees approaching and three females, who had possibly been following the progression of the hunters from not too far away, rushed to join them. The long search for the prey and the short hunt was followed by 2 hours and 40 minutes of meat eating, at the end of which most of the chimpanzees seemed satiated by the meal.

Chimpanzees are hunters. All long-term studies on chimpanzees with detailed knowledge of their behaviour revealed that they hunt for meat. In Taï, they have been seen to hunt about every 3 days throughout the year, with a clear peak during the rainy season from September to November, when they hunt at least once a day. On average, male hunters of Taï eat 236 grams of meat daily, which is the equivalent of a large steak. Females, on the other hand, eat about 80 grams of meat daily, with the higher ranking females eating more than the others. In East Africa, the chimpanzees from Gombe and Mahale in Tanzania and those of Ngogo in Uganda have been seen to hunt at least as often as the chimpanzees of Taï. The amount of meat eaten might possibly be somewhat less there because they kill more infant prey.

Figure 3.3   Schematic hunt showing the different hunting roles: driver (solid outline of box), chaser (dotted), bystander, blocker (dot-dashed) and ambusher (dashed) concentrating on colobus monkeys (shaded ellipses).

Successfully hunting small and swift monkeys high up in the thin branches of the highest trees of the forest requires some level of organization from the much heavier hunters. Taï male chimpanzees needed to develop sophisticated cooperative hunting tactics, which some view as some of the most elaborate examples of cooperation in the animal kingdom. How do they proceed? During three quarters of the 274 group hunts we followed, Taï chimpanzees performed four complementary hunting roles (see Figure 3.3). Briefly: the 'driver' initiates the hunt by slowly forcing the arboreal prey in a constant direction and follows it slowly from branch to branch in the tree; the 'blockers' are hunters who climb trees to prevent the prey from scattering in different directions and by their sheer presence prevent the monkeys from moving in that direction; the 'chasers' climb towards the prey and, by rapidly running after them, attempt capture; eventually, the 'ambushers' silently climb in front of the escaping monkey to block its flight and close the trap.[9] Hunting success increases with the number of hunters, so that hunts in which all the roles are executed are successful in eight out of 10 times. During such collaborative hunts, each hunter synchronizes and spatially coordinates his movements with those performed by others, and sometimes anticipates their future actions.

Taï males have been seen to perform most of the complementary roles and may even switch roles during a hunt, demonstrating a capacity for role reversal and perspective taking. If males were selfish, they would not assume the role of driver nor ambusher as both rarely make a capture, while captures are regularly achieved by individuals following the hunt from the ground – the 'bystanders'. Meat sharing after a successful hunt allocates to drivers about three times less meat than to the captors of the prey. Interestingly, the ambushers that anticipate the movements of the prey and the other hunters are given an equal amount of meat as the captors. To make hunting worthwhile in Taï chimpanzees, the meat sharing rules prevailing in the group guarantee the hunters a greater reward than that received by the non-hunters, and make a distinction between hunters in favour of those that made the most important contribution towards the success.

Selfish individuals should reasonably wait on the ground for the prey to fall or perform the ambusher role that guarantees more meat. As a consequence, group hunting would become rare. However, clearly this is not the case as Taï chimpanzees hunt about 250 times per year. On the other hand, a cooperative individual approach is more compatible with the observations made in Taï forest; individual hunters assess whatever role needs to be performed for the hunt to be successful and carry out whatever roles are needed independently of their short-term benefit. Just as in a team of soccer players, individuals react opportunistically to the situation in hand while taking into account the shared aim of the team. Some players, like the defenders and goalkeepers, will rarely score a goal, but the success of the team critically depends upon their contribution. Group hunting in chimpanzees is similar in that the synchronization of the different coordinated roles, role reversal and performance of less successful roles favour the realization of the joint goal. Thus, the group hunting behaviour of the Taï chimpanzees is a strongly cooperative enterprise based on the contribution of many individuals.

DANGEROUS COOPERATION IN WARFARE

If cooperation to hunt small monkeys is common, it is also systematically observed when facing the much more dangerous groups of strangers. I will discuss territorial defence in more detail in Chapter 5, but for now it is important to appreciate that territorial defence can be very dangerous as outnumbered males may be badly injured and even killed. Despite these extremely high costs, males regularly cooperate and participate in patrols deep into neighbouring territories. Similarly impressive, when a group of chimpanzees moves in the forest and suddenly hears strangers, individuals farther back will rush forward to join the leaders to defend the territory. Many times I have been in the forest with a group of males who, when they heard strangers and started to move silently towards them, would slow down and wait for others in the group to join them. Invariably, more males arrived after a while, and these newcomers were greeted by the first males before all of them, usually six or more, began to move together to surprise the strangers.

Patrolling males cooperate for hours as they look for the presence of strangers in the boundary areas. They often make deep incursions into unknown neighbouring territories and search as a tight team for signs of chimpanzee, all the time being vulnerable themselves to a surprise attack by strangers. One day, in 1989, I was in the north of the territory with a large group of females with infants and some males that were feeding in a valley. After having rested for a while, they started to move again, towards a ridge in the north, where I knew they would find and start to crack *Coula* nuts. Keeping an eye on the males, I noticed some tension between them and realized that, while the females climbed up into the trees to crack nuts, some males headed north. Darwin, Rousseau, Ulysse and Brutus moved silently as a close team for 200 metres. They sat in a line and waited without making a sound, looking to the north. After some 10 minutes, Macho and Kendo joined them. They immediately continued on to the top of the ridge, where they sat and listened again, whilst behind us we could still hear the sound of the females cracking nuts in the trees. Suddenly, hearing something that I could not, the males stood upright with their hair on end, and moved as a pack down the valley towards the north-east. Brutus, the oldest, stayed at the rear and the younger males regularly checked on him. Whenever he sat, the others stopped in a long extended line, waiting for him to move again. Ulysse led all the way, and within 20 minutes they had rapidly crossed the large valley. As they reached the top of the next ridge, they were at least 2 kilometres into the neighbouring territory and became more wary, walking close to one another in a single line. They all adopted an 'Indian walk' avoiding cracking dead twigs or branches by putting each foot down carefully and slowly. In comparison, I must have sounded like a heavy pachyderm and they looked at me whenever I stepped on dead branches. They sat on top of that next ridge and listened silently for a long time. Some 45 minutes later, Ulysse looked at Brutus and with a typical rocking of his body towards the north-east, they all silently started to move again. I still had not heard anything and only felt the increased tension between the males. After moving for a further 20 minutes on the other side of the valley, I finally heard the faint pounding of nut cracking. Brutus now speeded up and they all continued in that direction. Brutus stopped regularly to listen and the younger males in front did the same, waiting for him. By now, we were clearly getting very close and after another intense exchange of glances with Brutus, Ulysse and the others fanned out in a line in front of me to prepare themselves for the attack. Brutus was still at the back and they proceeded in this formation for some 30 metres. Then we heard the unhappy scream of a baby close to a nut-cracking sound – the strangers were still unaware of the danger! I saw Ulysse in front of me with his hair on end, rocking his body back and forth for 10 seconds, and then without a sound, he charged with Macho and Kendo on his left, Rousseau and Darwin on his right, and Brutus following. As they disappeared ahead of me, I heard the high-pitched screams of surprised chimpanzees accompanied by the aggressive barks of the attackers. I ran and saw that the attackers had surprised some females cracking nuts in front of a large windfall tree. They fled behind the log, followed by the attackers. Judging by the sound of the screaming, a large number of strangers was present. The attack did not last long and I saw Brutus, Ulysse and the others silently running away at full speed chased by many strangers barking aggressively at them. Some mothers supported their males and also barked.

The strangers chased the attackers for some 100 metres and returned towards their females, drumming loudly on trees. I tried to follow the attackers the best I could although I was panting and out of breath. They simply ran back into their own territory, clearly too outnumbered to resist this counter-attack. After 30 minutes of quick retreat, we entered the valley which constituted the northern limit of the North Group territory and they started for the first time since the attack to call and drum loudly. The females who had remained behind cracking nuts answered some 300 metres to the west. During the whole patrol, as was so typical, the males kept perfect cohesion, always checking to see where the others were and following the lead of Brutus, although he led from the back. Even during the retreat, they remained together. Not even at the height of the dramatic attacks did I see any individual running back on their own without waiting for all the members of the patrol. Cooperation during territorial fights is complete and done with precise coordination.

Thus, cooperation is a major aspect of the social life of male chimpanzees and it was observed daily in the Taï communities. Although some display of leadership was apparent, especially during territorial defence, to improve coordination and decision making in risky situations, participation in such cooperative actions was always voluntary and, except for some hunts, participants never failed to take part. Male chimpanzees seem very successful in making a distinction between their personal interest in the sexual domain where they are rivals, and their position as hunters and group members, where they are generous cooperators.

ALTRUISM AMONG MALE CHIMPANZEES

In chimpanzees, rivals are not only cooperators, they are also altruists. That competitors should cooperate where they would gain personally is understandable in a sense, but that they would be altruist in favour of a competitor is more puzzling. In general, altruism remains an enigma as it is difficult to explain its existence from the point of view of selfish individuals. By definition, an altruistic act profits the receiver but costs the provider. Where altruistic acts have been observed in different animal species, it has been proposed that in reality these acts were either beneficial to the actors as they are performed uniquely for close kin, such as workers in a beehive who forgo reproduction to work for their mother, the queen, and feed their sisters, or because such acts are reciprocated at a later time making them beneficial in the long run.[10]

Adoption has for a long time been a puzzle because of its altruistic nature. Chimpanzees have long been known to adopt young orphans but in the first instance, adoptions are made by older siblings. Infant chimpanzees mature slowly, being breastfed and in constant contact with their mother for the first 5 years of their lives. After the next sibling is born, the former remains with the mother for another 5 years, and only at about 10 years of age will the young chimpanzee start to leave her for short periods of time. Not surprisingly, infants suffer a lot when their mothers die. Generally, if that happens before weaning, they will not survive. I saw young orphans in such a situation completely lose any interest in life and letting themselves die. They may, however, be adopted by an older sibling if they have one, and one can see such orphans being carried around for weeks or months. Sadly, however, they become weaker and weaker and finally die. For the very young, the absence of milk is the obvious explanation, but for

4-year-old infants who have already started to eat solid food, it must be more than just subsistence that makes them lose the will to fight for life. They simply cannot make it without their mothers. If they have already been weaned, and are lucky enough to have an older sibling adopt them, they will survive. In some instances, the orphans may stay close to female friends of their mother and an adoption will occur whereby these females support the young ones when they are in difficulties and share food with them. It seems that the bond with a supporting adult member of the group is extremely important as such adopted orphans not only survive but develop physically normally. Orphans over 5 years of age that are not adopted might also survive, but their physical development will be delayed by 4–6 years. It is extremely affecting to see a 6-year-old orphan who is in reality 10 or 11 years old! Thus, once the maternal bond is broken, the best future for the young chimpanzee is in adoption.

One day, in 1985, I met Brutus, the alpha male of the North Group, followed by little Ali, just 5 years old, who had lost his mother, Awa, a few days before. Intrigued by this association, which I was soon to call 'adoption', I followed Brutus closely for weeks and watched him quickly display some aspects of typical maternal behaviour towards Ali. For example, the limited visibility in the Taï forest means that infants regularly lose their mother if she does not constantly check their whereabouts and wait for them. Brutus had already started, like a good mother, to wait for Ali each time he heard the young chimpanzee whimper. Moreover, I saw him retrace his steps so that Ali could see him, a behaviour I had never seen before in an adult male in all the years I had followed wild chimpanzees. But Brutus seemed willing to go much further in helping Ali. Brutus was not only the dominant male of the community, he was also the greatest hunter and constantly the proud possessor of large pieces of meat. He generously shared the meat with many different females and some males, but never with subadult individuals as usually it is their mothers who share with them. Since the adoption of Ali, however, Brutus also shared meat with him and this was the cause of constant squabbles, as the adult meat beggars did not approve of the favoured treatment shown to the youngster. This caused more conflicts in the community, but Brutus kept on sharing with Ali. He even handed over to him some of the most highly prized pieces, such as the brain of the prey.

Interestingly, as Ali followed Brutus all day long, he also followed him during the hunts, copying all the complicated movements to corner and block monkeys in the trees. Thanks to his very special 'mother', Ali quickly learned not to fear the monkeys and started to tease them and even faced adult monkeys without the loud screams of fear usual for 8-year-old youngsters. After a few years of this personal training, Ali showed some unusually early tendencies to hunt and initiate anticipatory hunting movements as he had seen Brutus do so often. Thus, at 12 years of age, Ali had acquired some of the cooperative hunting skills that are typical of 30-year-old chimpanzees.

Like a typical mother–juvenile pair, when Brutus was pounding nuts, he was seen to share pieces of nuts with Ali. They would sit facing each other on either side of the anvil. Using a heavy stone, Brutus would crack the very hard Panda nuts for hours and all the time Ali would be allowed to eat pieces of the kernels as he used to do with his mother Awa. For 8 years, Brutus was seen to share Panda nuts regularly with his adopted son. Four years into this adoption, we were able to perform a genetic analysis of both Ali and Brutus, and to my great surprise, Brutus did not seem to be Ali's father! The lack of self-interest seems obvious in this case.

Learning that Ali was not Brutus' son gives rise to the question of whether Brutus was aware that he was not Ali's father when he adopted him. As we know that chimpanzee males can recognize their own offspring, is adoption in chimpanzees more than just an empathic reaction to a youngster in distress?

Ali was only the first in a long series of adoptions by adult males that we saw in Taï. Brutus adopted a second orphan, Tarzan, 5 years after he had adopted Ali. We would then regularly see Brutus, still alpha male, walking in the forest with his two adopted sons. Brutus was even seen to offer his nipple to this 5-year-old youngster when Tarzan was stressed and needed to be reassured. At the same time, another male, Ulysse, also a great hunter, adopted another young male orphan. Like Brutus, Ulysse waited for the young one and shared meat with him. However, Ulysse was middle ranking and when the other males tried to challenge him, it was not unusual for them to harass the young chimpanzee. Such harassments quickly became too much for him and Ulysse put an end to this adoption.

More cases of adoption of orphan infants by adult males were observed in Taï. In the East Group, an adult male, Fredy, adopted a 3-year-old male, Carim, after his mother's death; in this case, he accepted him in his nest every night for 5 months! This adoption again surprised me, as it was the first time an adult male had demonstrated such clear maternal behaviour as sharing his nest with such a young one. Two further cases of adoptions of infants younger than 5 years of age were seen. In both cases, the males not only shared the nest with them but went as far as carrying them on their backs during the day. Especially enlightening was the case of the adult male Porthos, who was seen to carry Gia, a young female orphan, for months on his back. One of her hands had been injured in February 2007 and for a while she could not use it for walking. She complained a lot and Porthos started to carry her on his back. This led to an adoption when she was estimated to be only about 3 years old. Porthos responded like a mother to her whimpering, carrying her when climbing a tree, making a bridge with his body for her to cross the gap between two trees and signalling to her to climb on his back with the typical maternal movement of the arm with fingers outstretched when he wanted to continue walking. Gia's adoption by Porthos lasted for many months and Porthos carried this little girl on his back during some of the most dangerous male activities (see below for a dramatic example during a territorial fight).

All these males were competing with one another to sire more offspring and showed great interest in sexually active females. They were good hunters, investing a lot of time in attempts to capture monkeys, and were great warriors, sometimes taking huge risks in attacking intruders. But they were also willing to show great altruism and adopted infants in distress for months and years, even if they were not related to them.

LIFE-SAVING ALTRUISTIC SUPPORT

Altruists take even greater risks in chimpanzees. One day, I followed four adults of the North Group patrolling deep into the territory of their northern neighbours. They had left the community range some hours before and were silently covering a large area, not feeding at all, listening solely for the presence of neighbours. They regularly sat at vantage points and looked around. After 6 hours of this silent patrol, the males suddenly rushed ahead and loud calls resounded through the forest. I rushed and saw the four males surrounding a female neighbour with her baby. The males were slapping her on the back as she crouched on the ground protecting her baby. Whenever the aggressors paused, she faced them screaming, but was unable to resist or flee. When the attack was in its second minute, a second female suddenly broke out from the trees and, barking loudly, attacked the males. They immediately turned towards her and the first female escaped. The new female was quickly subdued by the males and had then to sustain the strikes and bites. Luckily for her, 2 minutes later, we heard the males of the neighbouring community approaching fast and drumming, which caused the four aggressors to immediately retreat. It is obvious that the second female took enormous risks to help her companion.

Porthos' altruism went as far as shouldering both the costs of adoption and of rescuing others! One day in February 2007, he was alone with his adoptive daughter, Gia, when he heard tremendous screams and without hesitation, taking Gia on his back, he rushed towards the calls. There, five males of the South Group had captured Bamu, a female of Porthos' community. Bamu, who was badly handicapped having lost an arm some years earlier, was an easy victim, being unable to resist these very aggressive males. Despite the overwhelming odds, Porthos on his own, with Gia on his back, charged the South Group males through the thick undergrowth. The loud aggressive calls of Porthos were, however, impressive enough to deter them and to save Bamu. It was breathtaking to see this adult male with a baby clinging to his back charging five male opponents. The intruders were a powerful and dangerous team as they were seen to kill another male of Porthos' community 30 minutes later (see Chapter 5). Porthos clearly risked his life but, fearlessly, did not hesitate a second to intervene to help a female of his community.

Altruistic support in the case of intercommunity violence is more than anecdotal, as it happened regularly over the 25 years that I followed these relationships. In almost a third of all visual encounters with neighbours ($N = 130$), some males rushed to support the individuals facing a difficult situation, and this proportion rose to more than three quarters when females of a community had been outnumbered and taken prisoner by neighbours. In some cases, a single male would not hesitate to rush in, while in others a group of males ran in together. In all instances, they used the element of surprise and remained silent during the whole approach, calling only just before being seen. Such supportive behaviour was always successful in freeing the females and spared them from the worst injuries.

Furthermore, male altruism was observed during leopard attacks. As mentioned previously, in the Taï forest, leopards have developed a special liking for chimpanzee meat and do not hesitate to attack them. Chimpanzees have long been thought to be free from predation because of their size and, therefore, able to organize their social structure without worrying about predation.[11] Our observations of leopard attacks on Taï chimpanzees changed this view; not only do leopards actively search for them but are able to kill adults without much difficulty.

I gained the impression that they prefer mature juvenile chimpanzees that walk independently in the forest but are not yet able to mount much resistance. On 8 March 1989, Grégoire, a project assistant, was following a male in the forest when he suddenly heard loud alarm barks ahead of him. His target chimpanzee ran immediately in rescue, and when Grégoire joined him again, he saw a chimpanzee lying on the ground and he hurried back to the camp to ask me to join him. I recognized Tina, a 10-year-old female, with punctures from a bite on the neck and long cut marks on her chest and belly with a small section of intestine protruding. She was dead. Tina was the older sister of Tarzan, the orphan male adopted by Brutus. At that time, Brutus was regularly seen walking in the forest following Ali, 10 years old by then, and followed by Tarzan with Tina lagging behind because she was wary of the big male. Her respect for Brutus probably proved fatal, since the leopard let this reconstituted family walk by and then attacked the last individual at the rear. Brutus turned back to defend her, but it was too late. The leopard broke her third vertebra with one bite and fled, possibly chased by Brutus. Within a few minutes, 15 chimpanzees had gathered around Tina and the adult males started to guard her body, imposing respect on all group members, especially the youngest ones who wanted to play with it, and waving away the increasingly abundant flies. The situation was perceptibly tense as if all were aware that something unusual had happened. Several times the guardians played in a rigid way for short periods with one another.[12] Only after 4 hours was the body left unattended for a few minutes and it was only after a full 6 hours that they finally left, when it was covered with thousands of flies. The leopard came back at night to eat both legs and the hips.

In a period of 5 years, the North Group was successfully attacked many times by leopards; successful in the sense that we ascertained that five individuals, including two adults, Salomé and Falstaff, were killed as well as the juveniles Hector, Tina and Homère, and nine were injured. All the injured victims of leopard attacks were tended by other members of the group. Caring included carefully licking the wounds and removing dirt from within the injuries. Immediately after an attack, the victims are often quiet, visibly in pain, and group members wait for them and repeatedly look after their wounds. An injured individual receives support for days and, in the case of awkwardly positioned wounds, such as on the back, head or neck, tending is extremely important. In the case of Falstaff's injuries, he received regular care from other adults until his death, 6 weeks later.

Leopard predation is life threatening to chimpanzees and support by community members can be vital for survival in a forest like Taï, where the density of leopards is high (about 7 individuals per 10 square kilometres). Male and female chimpanzees do not hesitate to rescue group members whenever they hear the alarm call of individuals in danger and, without faltering, rush in a tight group to support them. Immediate and speedy support is given to all attacked individuals. Leopards have extremely powerful jaws and support, if provided, has to arrive within seconds of the attack. Later is too late. As we saw in the previous chapter, Ondine and her female partners rushed immediately to support Ella who was fighting with a leopard and the aggressive barks of the supporting females certainly caused the leopard to flee, sparing Ella even worse injuries. Similarly, when Fossey was attacked, Schubert immediately rushed to the spot where she was calling and within seconds the leopard retreated.

Chimpanzees are generous in their support and ready to provide it in preventive situations. Another day, a scattered group of chimpanzees was moving in a flat area near a river with some dense patches of undergrowth. Suddenly, they spotted a leopard and Brutus barked aggressively, immediately supported by the others and, within seconds, seven adult chimpanzees chased the leopard, which tried to trick them by making abrupt changes in direction but the chimpanzees at the back of the pursuit cut short its turns each time. After 100 metres of the chase, the leopard took refuge in a hole in the ground under a large fallen tree where immediately the chimpanzees set up a siege. Many females and their infants came close to the hole and peered at the leopard, as if the mothers wanted to make sure their young learned to recognize this danger. At the entrance, Ondine joined in by using a long, about 4 metres, thick branch to stab the leopard. The leopard reacted right away by hitting the stick with its paw as close as possible to the chimpanzee's hand. Each time, the sitting males barked and the leopard retreated into the depths of the hole. For four more times, different chimpanzees, males and females, selected new branches and tried to stab or hit the leopard. After 3 hours and 45 minutes of siege, the chimpanzees left and the leopard sneaked out unharmed, 30 minutes later.

In other situations, chimpanzees have been seen to kill baby leopards. In three cases, they killed them by hitting the head against the ground. On one occasion, the chimpanzees surprised a mother with two cubs hidden under a fallen tree. The males immediately chased the mother away, while Zora, the oldest female of the South Group, caught the first cub by the tail and, swinging it around, hit its head repeatedly against large branches of the tree, quickly killing it. She killed the second cub in the same way, abandoning the bodies on the ground. The third killing of a cub was seen in the Middle Group, where the cub was killed by the bite of a male chimpanzee. Adult leopards are probably too powerful for chimpanzees to be able to do more than scare them away. Once, when they had caught a juvenile leopard, they hit it and threw it around a few times, but were not able to harm it seriously.

Male support in the case of leopards can be extremely risky and Falstaff most probably lost his life in this way. Falstaff was the oldest male of the North Group but still very active. He seemingly rushed to support a group member threatened by a leopard and when I arrived he was covered in blood and scars, very similar to the female Ella, who had rescued her small son (see Chapter 2). All the females and some of the males cared for him, licking away the blood and cleaning dirt from his wounds for hours. He would have survived, if the claw of the leopard had not perforated the lung under his right arm. This injury did not heal and he died 6 weeks later. Altruism in the case of support against leopard attacks can be deadly.

High-risk altruism towards non-relatives is a puzzle. It does not fit with our view of animals as selfish individuals. Nonetheless, we should recognize that altruistic behaviour is directed only towards group members, and the resulting within-group solidarity can in the long run profit everyone. In other words, be nice to others, they might be nice to you in the future. For example, Malibu was attacked in 1989 by a leopard and rescued in a strong response of the males, including Falstaff. Some months later, Malibu was one of the females carefully tending the injuries that Falstaff sustained from a leopard attack. Similarly, Macho was one of the males who reacted so quickly following the leopard attack on Ella, and later he profited from a similar strong male support when he was trapped by three stranger males and sustained 18 bleeding bites within 2 minutes. I could give more examples showing that in an environment with high predation pressure it makes sense to develop altruistic tendencies, as such support will over time profit many different group members including those that have themselves provided support previously.

FROM COMPETITORS TO ALTRUISTS

What is so special in the situation of the chimpanzee that new forms of cooperation and altruism should be observed? Like most social living animals, they have to face competition for food. Furthermore, as in most animals living in groups with more than one adult male, competition for sexual partners is observed. Both types of competition limit social cohesion. The level of male sexual competition in multimale social settings will be exacerbated in species with longer periods of maternal investment in the offspring, as this decreases the number of sexually active females. An increase in maternal investment is already visible in many primate species compared with other mammals, where maternal investment is mainly limited to suckling. In primates, mothers additionally often invest in the socialization of their infants. However, with the apes, we observe a further increase in maternal investment due to their slow maturing and longer living nature. Thus, one way for the males to increase their reproductive success is to increase access to females of other communities. This constitutes a strong incentive for otherwise competitors to cooperate in contests against groups of strangers.

Furthermore, chimpanzees in certain habitats face high predation pressure from leopards and such pressures lead to increased within-group solidarity. Anti-predator reactions are commonly seen in many animal species, but in most situations are equivalent to self-defence and often are performed jointly in groups. The mobbing of predators, for example, is observed in many species, ranging from small birds mobbing crows or raptors to monkeys mobbing eagles, or Gombe red colobus monkeys mobbing chimpanzees as they pass nearby. Predation pressure typically leads to a process of coevolution, whereby the quicker you run away from your pursuer, the quicker he will run. In other words, the level of predation pressure will influence the way you respond to it. The higher the predation pressure against large social animals, the higher the costs for the group members, and the more likely they are going to react to such attacks. The high density of leopards found in the Taï forest places the chimpanzees in such a situation, where remarkable effective responses are required. Taï chimpanzees systematically mob leopards whenever they see or hear them and they consistently rush to support group members during attacks and provide extensive care for victims of such attacks. Victims' survival is decisively improved by the care provided to them by group members.

What distinguishes chimpanzees from other primates here is that they might have realized that high social fraternity, in terms of cooperation and altruism, when facing dangerous enemies, like leopards and neighbours, will permit group members to fare better. This realization requires empathy and compassion – in other words, the ability to understand that another individual is in need of help, and that you as an independent individual can provide this and as a consequence the one in need will be doing better. If for humans this ability may sometimes be obvious, it is very rarely seen in other animal species. Therefore, in my view, the reaction of chimpanzees to predation pressure by providing care and help to other individuals is rooted in their more developed social understanding. This ability to appreciate the perspective of others seems less developed in monkeys than in chimpanzees. For instance, consolation by third parties after a conflict between two opponents has been recorded only in chimpanzees and has never been reported in baboons or macaques. Self-recognition in a mirror has been observed only in apes and not in monkeys.[13] Therefore, I suggest that chimpanzees but not monkeys have the basic cognitive ability, namely empathy, that allows them to be able to cooperate *extensively* and help others when needed.

One key area where such intelligent abilities are of a central importance is when hunting. The more hunters can predict the behaviour of their prey, the more successful they are going to be. Specialized hunters, like hyenas, have been shown to possess a developed understanding of others, allowing them to cooperate.[14] This understanding and predicting of others including prey would be even more important when hunting is done in a low visibility environment, like the rainforest. This might explain why in both humans and chimpanzees the most difficult hunting roles are performed only by the older individuals that have about 20 years of experience in hunting.[15] Thus, because chimpanzees are good cooperative hunters, they have the basic ability to consider the need of others and, when facing high enough predation pressure, will develop cooperative and altruistic behaviours to mitigate the effects of predators.

Thus, the combined effects of high sexual competition on one side, and of high predation pressure on the other side, will produce the mixture of high within-group solidarity with high between-group aggression. Once such behaviour patterns have been acquired, it would be easy to generalize to other social situations. This is especially true for the adoption of weak group members, like orphan infants. In addition, strong within-group solidarity also leads to xenophobia. Male chimpanzees have this duality of high level of violence and high level of cooperation that allows them to support group members in bad situations, while in other contexts they are fighting for months to gain social dominance and more mating partners. We will see in Chapter 5 some of the complexities we observed in intergroup conflicts in Taï chimpanzees and how developed within-group cooperation can be and the possible consequences of a xenophobic tendency in chimpanzees.

NOTES

1. Penis morphology is extremely varied, both in length and form. The length of the chimpanzee's penis seems to be adapted to the increased length required for placing ejaculates advantageously to ensure adequate sperm transport within a female during the period of maximal sexual swellings, when the vaginal area is increased by about 5 centimetres (Dixson 2002). Other species have penises with different shapes at the tip or have spines which enable the raking out of competitors' sperm. For example, the human penis with its enlarged head seems to function as a sperm removal device and this is also seen in other species (Dixson 1998).

2. Testes have been observed to vary vastly in size in many animal species including the primates (Dixson 1998). Some variation is related to body size and phylogenetic effect but after allowing for those effects, the variation remains large. In primates, testis size correlates to the dominant mating system, with smaller testes in males of monogamous species and those in a harem system, and larger size testes in those belonging to multi-male groups or dispersed species.

3. Roman Wittig spent 3 years in the Taï forest documenting in great detail conflicts between the chimpanzees and observed how a peaceful social life is possible despite the many conflicts, and how fascinating 'peace-making' strategies allowed individuals to restore harmony after fights (Wittig and Boesch 2003a, 2003b, 2005).

4. Recent progress in detecting small amounts of hormone metabolites in urine and faeces has opened the way to hormone studies in wild animal populations. The most recent progress concerns studies of male hormones and gives the first indications of the levels of stress and aggression that individuals have to face in their natural social life (Muller and Wrangham 2003, M. Hauser *et al.* unpublished data).

5. Kin selection is a major tenet of the theory of social evolution, as we expect genetically closely related individuals to support one another more than if they are not related. In other words, individuals are not only expected to invest in their own infants but also in close kin and that investment should be proportional to the degree of kinship (Hamilton 1964). Since wild animals are expected to be living in groups with many kin, kinship was predicted to contribute to the evolution of many forms of social behaviours. The major difficulty with this important theoretical idea is that it rests on the assumption that wild animals have the ability to recognize kin and different levels of relatedness. Evidence of kin recognition has been difficult to collect, mainly because it has been hard to distinguish between real kin recognition and preferred associations with familiar individuals, such as would be expected between individuals that have grown up together. Individuals that associated more were more prone to act together in general. Nevertheless, ample evidence of kin recognition and kin-biased behaviour has been documented in wild social animals (Clutton-Brock 1991).

6. Infants start to play when they gain more independence from their mothers and that varies between 18 months and 2 years of age. Grooming, on the other

hand, develops later in youngsters and becomes common with the birth of younger siblings, and juveniles use grooming as a way to maintain close contact with their mothers. Thus, it should not be a surprise that paternal investment appears at different ages for different behaviour patterns (Lehmann *et al.* 2006). Offspring recognition was expressed clearly in the behaviour of the father in terms of play behaviour with offspring and decrease of aggression against the mother of the baby. Our results with the Taï chimpanzees are not the first in primates. The first case was shown in savannah baboons, where males more readily supported their own offspring in conflicts against other males (Buchan *et al.* 2003). In this species, males form strong consorts with females during their reproductive period and they seemed to be using this to predict their paternity.

7. Cooperation between unrelated individuals remains a central challenge for biologists as it is difficult to explain why individuals should invest in joint activities when success is not guaranteed and when cheaters can obtain some benefit without investing (Dugatkin 1997, Axelrod and Hamilton 1981, Maynard Smith 1982). No matter how you look at it, if cheaters have equal access to the benefits, then cheaters do better. Thus, there is a need for a mechanism that selectively rewards the individuals investing in the cooperative act rather than those that do not. If you live in a group of highly related individuals, this is not as important as the cheaters are your kin (Nowak and Sigmund 1992). If not, however, then enforcing such a mechanism is cognitively demanding and it has not been found often (Boesch and Boesch 1989, Boesch and Boesch-Achermann 2000).

8. Cooperation in lions has long been the best known example of cooperation in animals. Since George Schaller's observations of Serengeti lions (1972), we know that groups of lions capture more prey than individuals. However, more detailed observations have shown that individuals gain less meat when hunting in groups than when alone (Packer *et al.* 1990, Grinnell *et al.* 1995). At the behavioural level, they also observed that only a few individuals in such a group really hunt, while all try to get the meat. Lionesses are in a group to improve the survival of their young and when together they also hunt, but they do not seem to live together to hunt more efficiently.

9. Taï chimpanzees hunt about once a day during the 3 months of the rainy season and about once a week during other months. Group hunts are observed in 84% of hunts, of which 77% demonstrated the elaborate division of hunting roles (Boesch 1994, 2002, 2005; Boesch and Boesch-Achermann 2000). Selection of prey is very strong in the Taï forest and 95% of the prey is made up of colobus monkeys, despite the abundant presence of small forest antelopes and 10 species of monkeys. The purposeful nature of the hunt is illustrated from the fact that in half of the hunts chimpanzees had been seen searching for the prey for up 2 hours.

10. Altruism in animals has often been denied either because it is theoretically difficult to explain when starting from the point of view of selfish individuals (Axelrod and Hamilton 1981, Maynard Smith 1982), or because it has been proposed that animals are unable to have empathy for others, a cognitive capacity that requires an understanding of the distress of another individual. For some, only humans are proposed to feel empathy, and so altruism should only be expected in humans (Tomasello and Call 1997, Silk *et al.* 2005). Animals, however, are known to take greater risks when helping close kin and, therefore, kinship is expected to lead to altruistic acts. In addition, reciprocal altruism has been proposed by Trivers (1971) as a solution to this dilemma in the sense that individuals would be willing to incur

costs in favour of another, as long as there is a reasonable probability that this act might be repaid in the future. Obviously, such a situation would be more likely to occur in long-lived animal species living in stable close-knit social groups. Chimpanzees and other great apes live in conditions that mirror such a situation.

However, in an experiment, captive chimpanzees were shown to be indiscriminate in choosing either an action that benefits only themselves or one that would benefit both themselves and a companion. From this it was concluded that 'Chimpanzees are indifferent to the welfare of unrelated group members' (Silk *et al.* 2005). New experiments carried out with other individuals and with another apparatus showed exactly the opposite, namely that chimpanzees are spontaneously altruistic with humans (Warneken *et al.* 2006).

11. Prey animals normally have relatively few weapons with which to defend themselves compared with the weapons of the predators, so that the easiest defence is simply to hide within a group of conspecifics, where the likelihood of being killed by a predator decreases proportionally to the number of individuals in the group. However, maintaining cohesion of such groups in a low-visibility environment requires vocalizations, which will attract predators. In the Taï forest, we observed many instances where leopards approached chimpanzees, following their vocalizations, and were successful in surprise attacks (that was the case in the attack against Tina and Ella).

12. The notion of death is generally attributed only to humans; however, unusual behaviour related to death has been observed in a few species. Elephants are known to revisit sites of previous deaths and seem to recognize the bones of dead parents (Moss 1988, Poole 1996). Taï chimpanzees have extremely intriguing behaviour in this context, not only imposing respect for the dead body on others but treating blood on the body differently from when the victim is injured, that is, licking and cleaning wounds, rather than touching a dead body (Boesch and Boesch-Achermann 2000).

13. Many studies on post-conflict interactions in mammals have shown that if reconciliation between former opponents is seen in all species, consolation whereby a party not involved in the conflict brings together the two opponents is seen only in chimpanzees (de Waal 1993, Wittig and Boesch 2005). This is in parallel with the lack of self-recognition in mirror experiments in all monkey species tested (Gallup *et al.* 1971, Eddy *et al.* 1996). Hedwige Boesch and I suggested that it is the fact that all wild chimpanzees hunt prey for meat and by doing so are forced to consider and anticipate the reaction of another species if they want to be successful (Boesch and Boesch-Achermann 2000). To become successful hunters, chimpanzees seem capable of learning the perspective of others.

14. An experimental study with captive hyenas showed that they have the ability to take into account what another individual does before pulling on ropes to get a food reward. The reward is obtainable only if two individuals pull at different ends of the same rope (Drea and Frank 2003).

15. The Ache of Paraguay achieve the best performance in hunting when about 40 years old, although they start to hunt at 17 years (Hill 2002). Similarly, Taï chimpanzees perform the more difficult ambusher role once they are 30 years old (Boesch and Boesch-Achermann 2000). In both cases, the difficulties in anticipating the actions of the prey in poor visibility were suggested to be the basis of such a long apprenticeship.

4

Odyssey through our forest past

Poupée and Malibu as well as Brutus and Falstaff have grown up in the depths of the dense tropical rainforest of Côte d'Ivoire. Their life horizon has always been limited by trees, with a few openings made by treefalls, while the rivers cutting through their range are too small to open up the forest. This kind of forest is predominantly flat, so that even when they climb in the highest trees, the visibility is restricted to the neighbouring trees and the sky. Their home is a large forest tract of over 4 by 5 kilometres that they criss-cross on a weekly basis, and they tend to join the seasonally fruiting trees in straight lines even when these are quite far apart. How do they keep contact with friends and partners in such a forest? How do they know about them, their where-abouts, about potential danger, about all that matters in their social life? Being a social animal in a dense forest presents some very special challenges. Many animal species reacted to such conditions by becoming less social; the lovely red bushpigs with their brushy white ears live in much smaller groups in the forest than in more open habitat, the small forest elephants are mostly solitary compared with the large herds of their savannah counterparts, while the buffaloes are encountered only in tiny groups in the forest. Forest leopards are solitary except for the short mating periods. This contrasts dramatically with the many social carnivores living in the savannah. How do chimpanzees manage to remain so highly social in the forest?

FROM DAWN TO DUSK IN THE FOREST

Early one morning, I managed to catch up with Salomé whose fertile period was visibly culminating as all the males were in a craze, following her through the woods, the swamp, up and down trees. Whenever she stopped, some would immediately approach her, to be chased away just as immediately by other pretenders. However, when Schubert, the second top-ranking male of the community, came near her this was more serious and Brutus, the boss, intervened personally. Tension heightened rapidly between the two dominant males, both trying to impress one another. This seemed Falstaff's time, the oldest, former high- and now low-ranking male and Salomé's years-long faithful fan and supporter in tense social situations. He tempted his chance during the few moments when the two dominant ones were busy challenging each other and did not control Salomé. Her oldest son and Falstaff's young follower, 17-year-old Snoopy, were also present, but his fascination was rather for the high tension between the males than sex, and he was never seen to try to mate with his mother. Despite this chaos, Brutus remained the only one I witnessed to mate with Salomé on the ground, in full view of all the other males.

Hearing red colobus monkeys calling some 100 metres away, Salomé led the males in that direction – it seemed she felt like a bite of meat. When they arrived silently under the tree in which the monkeys were gathered, her 4-year-old son Sartre, a keen young hunter, cautiously started to climb while Falstaff got into position and carefully watched the reaction of the colobus monkeys so as to be able to anticipate their flight movement. Correctly anticipating the direction, he hurried and climbed the tree to which the monkeys were heading, followed by Brutus who never turned down an opportunity to hunt. Taking advantage of his distracted rival, Schubert mated with Salomé, which caused Brutus to rush down and interrupt the mating, displaying impressively towards the pair, and chasing the screaming Salomé round and round. Schubert, out of sheer frustration, displayed towards some females and, following the general trend, Falstaff also displayed wildly. The forest resounded with the screams and barks of the females and the drumming of displaying males around the highly desirable Salomé. In the end, no one mated with her and the red colobus, now totally alerted to the presence of the chimpanzees, scattered among the trees.

Once the calm had settled, Salomé led the whole group towards the calls of frogs ahead in a valley where large fruit-laden *Sacoglottis gabonensis* trees grew. The ground under the trees was covered with masses of hard prune-shaped fruits and a delicious smell of marzipan floated in the air. Brutus displayed again and a group of 10 chimpanzees with enthusiastic food grunts started collecting handfuls of fruit, sniffing them to assess their ripeness. Salomé, too, happily gathered fruit, walking upright, followed by Sartre. She then sat at the edge of a large puddle of water into which she emptied her collection of about

15 fruits. After rolling them around in the water to remove the earth, she peeled four of them, one by one, with her teeth and chewed them into a wad of pulp, pressing it hard in her mouth, in between repeatedly dipping it into the water, to suck out most of the sweet juice. All the other chimpanzees were sitting beside the many puddles under the trees, busily chewing and dipping. Once all the juice was pressed out of the wad of pulp, leaving on it a clear imprint of the jaw and teeth, it was left by the puddle and the whole process restarted with fresh fruits. Lazy Sartre begged with an outstretched hand for his mother's wad. Salomé rapidly cut it in half and handed part of it to him, and he then happily dipped and chewed it.

After 20 minutes, they started to move downstream from one *Sacoglottis* tree to another, collecting fruits and carrying them to puddles of water. The forest was littered with the remains of chewed pulp that turned red within minutes because of oxidation in the fresh air. They foraged calmly for some 90 minutes, moving about 200 metres down the river. Once, Brutus mated with Salomé in full view of three other males, and Schubert displayed for a short time, but away from the couple. Suddenly, they heard the alarm calls of a squirrel, which they answered by making intrigued hoohs for about 3 minutes. There are different reasons for these alarm calls, one being the sight of human observers, but another is particularly important for chimpanzees – the presence of a leopard. Squirrels, when spotting a leopard, give long alarm calls and, remaining in the trees, approach and follow the cat for as long as it is in the area. Chimpanzees always react to these alarm calls, and if the calls continue, they will even make charges in the direction they are coming from. But, this time, it did not seem too serious and they continued processing their fruits.

Salomé then found a large windfallen tree and sitting on top in the daylight started to groom Sartre. The other males joined in and Brutus groomed Salomé, while being groomed by Falstaff. Brutus stood presenting his buttocks to Falstaff, while Salomé held her face motionless towards Brutus. Brutus held her head with his left hand and with the right one carefully removed a spot from the corner of her eye, a gesture chimpanzees really seem to like, while Falstaff lifted Brutus' left testis and, with noisy and enthusiastic tongue clicking, removed a tick or something similar. He placed it on the inside of his left forearm, an almost hairless spot, and started hitting the tick fast and repeatedly with the index finger of his right hand. Brutus, Salomé and Schubert had rushed forward to look closely at what Falstaff was doing. Twice, he put the tick in his mouth, manipulating it between his lips and tongue, and then returned it to the forearm to hit it again, while the other three looked on with great concentration. Finally, he kept the tick in his mouth and ate it, and they all resumed their grooming session. Then they all stared towards the undergrowth where something rustled. Ondine, with her daughter Orée on her back, joined them with a small grunt to greet Brutus. Salomé immediately went to meet her best friend. They both sniffed at one another and Ondine showed her back to Salomé. Both sat, and Salomé started grooming her friend. The males placed themselves around the two females and the grooming session resumed. Ondine is the highest ranking female of the group and Salomé's almost constant associate. However, the chaotic excitement around Salomé's sexual swelling was too much for her, and during this period, they were less frequently associated. However, Ondine always seemed to know where Salomé was and joined her whenever the situation was quieter.

Then the groomers changed positions and Brutus, Falstaff and, for a little while, Schubert all groomed Salomé and Ondine, while Brutus and Falstaff also groomed each other. Grooming is a very regular and serious activity in chimpanzees that helps to remove the ectoparasites that settle on all parts of the body, especially those not directly accessible by the individual. At the same time, it is a close and intimate activity as the neck, armpits, buttocks, testes or sexual skin are presented to others, which is a very trusting act. It is an important social activity performed with social partners or carried out to enlist support and tolerance, and it is not surprising that chimpanzees invest a lot of time in it.

Suddenly, a drumming was heard far away to the west, which elicited loud pant-hoots and short reassurance-seeking behaviour by touching each other. Then they ran off through the forest in that direction with Brutus drumming and hooting powerfully on the first thin buttressed tree that he encountered. Another group centred on Macho and Ulysse drummed as well from a short distance. Recognizing Brutus (at least as well as I did), Macho immediately drummed again while a third group, females this time, joined in the chorus from another direction. Group cohesion in such low visibility forest is primarily maintained by calls and drumming. The voices of the chimpanzees are easily distinguishable and after this outburst of calls everyone knew the whereabouts of most group members. Even human observers can recognize chimpanzee voices within seconds, admittedly with some training. So, chimpanzees keep track of each other's movements even when out of sight and they identify those who approach from any given direction. It can be crucial to remember where the others are, especially your allies, as they might be needed suddenly when facing social squabbles or attacks from enemies. Similarly, dominant males have a habit of approaching by charging directly from a distance and it may be wise to avoid them before they can charge you aggressively. Furthermore, precise knowledge of individual voices of all group members allows the recognition of strangers and therefore a speedy reaction to any intrusions into the territory.

They moved rapidly as a close group in the direction of the first calls. I remembered the early days as we were struggling to keep up with the chimpanzees. If chimpanzees can be the noisiest animal in the forest, screaming at the top of their thunderous voices without inhibition, pant-hooting, barking and drumming for long moments, they can also, the second after, become totally silent and remain so for hours, as if they had just vanished out of reality. Such elusive traits make sense as predators are lurking around all the time, but it can make the life of scientists extremely difficult and considerably raise their level of frustration – afraid they might not be far away, just performing fascinating new behaviours, but being utterly unable to locate them. Following chimpanzees in the forest is a skill not taught in university classes and all researchers have to be lucky enough to work with animals that help them to overcome the handicap of being a slow bipedal. In my case, it was Falstaff. One of the oldest males in the group, he was mostly at the rear of the chimpanzee group I was trying to follow and therefore we spent a lot of time together. I am not saying he liked me, but this persistent clumsy and noisy follower seemed to intrigue him: why does he make so much noise when crawling through a thicket? Why does he fall so often in the wet slippery rivulets? Why can he not see me as he struggles out from under a treefall? Especially in those last instances, Falstaff would lose patience and throw sticks at me if I did not notice him quickly enough. Despite being quite old and often lagging behind, Falstaff was still a healthy adult male chimpanzee, and I either had to learn fast or lose this opportunity to learn from him the basics about moving in the forest. I made use of his lessons and soon understood some of their ways of foraging.

While moving, Salomé apparently smelt something and stopped, looking on the ground around her. The males tracked back and searched with her. Salomé and Brutus followed something on the ground – a trail of black driver ants! After following it for 30 metres, they arrived at the nest, the location of which was betrayed by a mass of loose soil around a tree trunk. Immediately, Brutus raked away the soil from the entrance to the nest, which was being aggressively guarded by numerous soldiers. In a fast move, Brutus inserted his arm right up to the shoulder into the nest, while Salomé and Falstaff, positioned clockwise around the tree base, did the same. Brutus removed a handful of grubs from the nest, many dropping as he stuffed them into his mouth. He repeated this with two more handfuls, then hastily ran 5 metres away, waving his hands about, hitting them against tree trunks in what looked like a mad dance. The soldier ants had started to bite and he was doing his best to get rid of them. He sat quietly and, taking his time, spat out the grubs into his hand and ate them with delight bit by bit, mixed with fresh leaves from a nearby *Coula edulis* sapling – something chimpanzees always do with highly favoured food. Salomé and Falstaff each had three handfuls of grubs and had also settled to eat, while Sartre looked on in fright and fascination at the masses of ants now in a full alarm state swarming out, climbing and biting everything close to the nest entrance. He held a stick with which to fish out some ants but, before he could even extend his hand, he was bitten and ran away dropping the stick. Sartre had got it wrong. It is the red driver ants that chimpanzees fish for with sticks. They dip the stick into the entrance and, in defence the soldiers bite and hold on to it. Chimpanzees then bring the stick to their mouth chewing energetically before the condemned ants can bite. There is no good reason, to my knowledge, not to fish for black ants as well; but it is simply not the way Taï chimpanzees deal with ants!

The desirable Salomé, still followed by her string of suitors, now climbed a tree to eat the hard pods of *Sterculia rhinopetala*, which contain very tasty seeds. These pods are extremely hard to open and Sartre, with his small canine teeth, was not yet able to open them, and so he begged for some from his mother, as youngsters commonly do with nuts cracked with hammers by their mothers. Of course, Salomé generously shared the opened pods, having struggled to enlarge a hole and extract the six to eight seeds from inside. While observing both of them in the tree, I heard the sound of a branch being banged on a root behind me. I turned and saw Ondine, Salomé's best friend, who had approached from behind me. She was sitting and splitting a hard dead branch in which there was a nest of wood-boring bees. She pounded with her fist at an adult bee on the ground as it tried to fly away and ate it quickly before it could sting. She then turned and cut a leafy branch from a sapling, removed the leafy part with her teeth to produce a stick about 18 centimetres long, and chewed some of the bark away from one end – making a tool within 30 seconds! Holding the stick between her thumb and index finger, she inserted it in a channel in the branch made by the bees and extracted the honey and larvae lump by lump. All the while, her tiny daughter Orée watched intently and sniffed her tool. Then, Ondine disappeared in the forest.

Ten minutes later, I heard the deep resounding pounding sounds made by cracking the very hard *Panda* nuts some hundred metres away. I left Salomé and her suitors to look for the nut cracker. First, peering through the vegetation, I saw the large movement of a chimpanzee lifting a heavy stone above the head to hit the nut placed on a root anvil. It was Ondine working at one of the identified *Panda* trees within the territory, a tree with its load of fruits now covering the ground. There had been no stone there before when I had checked recently. Thus, Ondine must have carried this 5 kilogram tool from its previous location. In fact, I then inspected the stone and could certify that it was the one from another *Panda* tree some 125 metres to the north and, of course, Ondine knew about it too and had gone for it! She was using her two hands to lift the stone and once the nut was open, Orée sitting in front of the anvil stretched her hand to obtain a piece. Ondine generously split a part of the whole nut for Orée to take. Slowly darkness fell and I could not prevent myself thinking about Héra who had lost her baby as a consequence of an attack by a leopard when she was on her own. Ondine did not in any way give the impression of feeling insecure and simply continued to feed on nuts. Just like me, a leopard could localize these nut-pounding sounds and know who was making them. I had often been wondering about such isolated mother and child pairs, potentially easy victims for an ambush killer. Does a leopard know that a chimpanzee with a hammer might present too high a risk as it turns against the aggressor?

We heard Falstaff still in the *Sterculia* tree giving his first night calls, signalling that he was ready to build his nest. Ten minutes later, Ondine stopped cracking the nuts and moved on for some metres as Macho and his group not far away also gave night calls. I had not seen Macho during the whole day but had heard him regularly communicating with others while moving along. Now another group of females called further away and both Macho's and Brutus' groups answered. The forest then turned quiet and after ten more minutes Ondine moved towards Salomé's group and joined them with Brutus and Falstaff also climbing adjacent trees to make their night nests. A nest is a platform of intertwined branches placed on top of two larger branches with a lot of leaves for comfort. Sartre joined his mother as he shares her nest until the next baby is born. With a last exchange of calls between the various parties in different trees, the last individuals finished making their nest and went to sleep. It was now almost dark and I still had a 45-minute walk through the forest to find my way back home, to eat and rest as well.

Eight months later, on 4 February 1988, Salomé gave birth to little Simone. Who was the father – great Brutus, Schubert, the young challenger, or Falstaff, her faithful friend of many years? Sadly, Salomé and Simone both died during the summer a year later before we could test them genetically for paternity. Simone's father remains a mystery just as many aspects of the life of the chimpanzees will always remain a mystery, leaving room for legends and the magic of imagination around our cousin in the forest.

The home of the chimpanzees, the African forests that covered the continent for most of its existence, is impressively rich in biodiversity. In the Taï forest in Côte d'Ivoire and the forests of Central Africa, where I work, tree diversity is very high with over 170 large tree species all producing fruits that are potential food sources for the abundant fauna living there. Besides the numerous birds and small forest antelopes, one major difference with more northerly forests is the large community of primate species. Eleven species of primates live in the Taï forest, including two nocturnal species. Eight species of monkeys, which are active during the day, forage noisily in the trees, including the Diana monkeys, also known as the 'captain' due to the obvious white stripe on their thighs. These give frequent alarm calls to warn group members of the presence of predators, mainly, in their case, the large-crowned hawk eagle. Diana monkeys are always found foraging together with other monkeys and these multispecies associations are unique to tropical African forests. The Diana monkeys with the other two guenons search constantly for insects to eat, while the larger red and black-and-white colobus monkeys concentrate on eating leaves. Red colobus monkeys are the most numerous in this forest and form large multimale and multifemale groups that are very active early in the morning with the males fighting each other over access to females in oestrus.

When walking through the Taï forest, it is not uncommon to encounter 10 of these associations of monkeys, including three to four different species every day. Their calls are heard frequently and there are estimated to be about 190 individuals per square kilometre, one of the largest concentrations of primates on the planet. Not surprisingly, crowned eagles, leopards, chimpanzees and pythons are known to prey on them. The presence of these large predators as well as smaller cats, like the golden cat, numerous snakes, like the cobra, mamba and

Gabon viper, as well as many more birds of prey, is a reflection of the rich biodiversity of these forests.

AFRICAN FORESTS AND OUR ORIGINS

For millennia, most parts of Africa have been covered by tropical rainforest, and the most important remains today are found in the tropical belt stretching from west to east. East Africa became the dry region it is today when this part of the continent rose above sea level millions of years ago. All the forested regions were probably inhabited by the same fauna that we see today in most of the remaining forests. I say 'probably', as it is difficult to find fossils of ancient times in forest regions because of the prevailing humidity. Nevertheless, we know from evolutionary studies that old world monkeys and the great apes originated in forested regions.

The once continent-wide forests have tracked the climatic changes of the continent. They became less widespread when the humidity decreased. This was for the most part influenced by the different periods of glaciation, the last of which ended about 13,000 years ago, lasted for a long time and was associated with high levels of speciation and divergence between the surviving animal populations. This is still apparent in many of today's forest monkey species. For example, the Diana monkey in Côte d'Ivoire is divided into three subspecies, their distribution delimited by the large rivers that still divide the forest into discrete sections: the forest west of the Sassandra river corresponds to the main forest refuge during the last glacial period in the western part of West Africa, while the second forest refuge was located to the east by the Comoé river in present-day Ghana. Similar sequences of divergence have been found in red colobus monkeys in West African forests.

The large gap in forest cover between West and Central Africa located at the Niger Gap has a very ancient history. It seems to coincide with possibly the oldest divergence in chimpanzees between the western populations located from Senegal to Ghana and the remaining chimpanzee population, east from the Niger River towards Nigeria and Cameroon.[1]

Africa is our continent of origin. The great apes evolved in the forest – gorillas in the Central African forests, bonobos in the forests of Congo and chimpanzees throughout the tropical forest belt from West to East Africa. Our first ancestors, *Sahelanthropus tchadensis*, *Orrorin tugenensis* and *Ardipithecus ramidus*, were discovered near or in the same African regions.[2] Similarly, *Australopithecus anamensis* and *A. afarensis*, who were closer to us, are now known to have lived in relatively wooded habitats. Since then we have evolved, but we carried with us the baggage inherited from our ancestors. This heritage imposes limits and directions on possible changes and adaptations. The elapse of time since our separation from the chimpanzee line is still unclear but might be much less than commonly thought. As humans, we tend to think that we are very different from others and, therefore, favour older divergence time – 18 million years ago was still being suggested in 1950, while some of the most recently discovered fossils suggest a divergence at some 7 million years ago. Whatever that time might be, it is clear that we share much with chimpanzees in terms of intelligence, sociality, cooperation and tool use, to mention just a few examples. Recent fascinating genetic studies, using information available since the successful completion of the human and chimpanzee genome project in 2006, suggest a much more recent and complex separation between us and our cousins from the forest. Complete separation between the chimpanzee and human lines might have been accomplished only about 1.4 million years ago.[3]

WHAT MAKES THEM CHIMPANZEES?

I hope that after reading the previous chapters, I have awakened the reader's interest in the chimpanzees. If that is so I would like now to provide some background information about them to explain the value of studying chimpanzees. Chimpanzees are not only fascinating by themselves; they are very interesting representatives of the life in the forest and our closest living relatives, which as such provide us with a look into the human prehistoric roots.

All the most recent genetic, morphological and behavioural studies concur in concluding that humans, not gorillas, are the closest living relative of chimpanzees. In other words, the term great ape, or 'pongid' to give its scientific equivalent, does not represent a biological reality. It is an artificial construct to support the notion that humans are a distinct entity in themselves, something which is not supported by scientific fact. Genetic studies agree in that the difference between humans and chimpanzees is very small at about a level of 1.5%. The morphological differences would be somewhat greater, which is exactly what would be expected as morphology reflects not only the genes but also the specific adaptations to the environment.

Some aspects of chimpanzee life seem to be consistent in all known populations. The first is social structure: chimpanzees live in stable social communities with many adult males and females. This is strikingly different from the gorillas, which are mostly found in one-male groups, a social structure that is common throughout the primate family, as in forest guenons and in two well-studied baboon species, the gelada and hamadryas. Monogamy is not common and is found only in the gibbons and some small New World monkeys. Orang-utans have a dispersed social structure that remains somewhat of a puzzle with individual territories of females encompassed in those of large males, while a few smaller males roam more or less freely in between. Bonobos, many macaque species and savannah baboons live like the chimpanzees in multimale, multifemale groups. However, community members in chimpanzees and bonobos are rarely all seen together and the flexibility of this fission–fusion system is very high in chimpanzees. In Taï chimpanzees, the parties we followed in the forest contained on average 13% of all community members and party composition changed frequently with individuals joining or leaving every 24 minutes.[4] So it is a very fluid society where most members only see a small number of their community at any one time.

Another important aspect of the sociality of chimpanzees is that, unlike most primate species, the males remain in their natal community all their lives, while the females transfer just before maturity. In savannah baboons, which have the same group structure, the males transfer at maturity and have to fight their way into new groups, while the females develop very stable social networks based on matrilines, with daughters often acquiring a social dominance rank next to their mothers. It is very different in chimpanzees, where the males develop intimate bonds while females are generally hostile at first to new immigrants. However, chimpanzees are long-lived and in the wild they can reach 50 years of age, so females can also develop close bonds with one another. This social system where the male remains at home is also the dominant system in humans and it was traditionally presented as a male-bonded system. Males remaining in their natal groups are more likely to live with brothers and cousins than if they transferred. However, since chimpanzee mothers transfer between groups before reproducing, this has a limited effect.[5] The transfer of individuals between social groups before maturity decreases the risk of inbreeding but when only one sex transfers, which is often the case in long-lived species, there remains a certain risk. In chimpanzees, sons may be attracted to their mothers. This is not generally the case and incest is avoided, but we know of cases where sons have mated successfully with their mothers and often brothers may be attracted by their sisters before they transfer. However, since young females have a period of sterility before and after transfer, such mating has limited consequences.[6]

Chimpanzees are long-lived primates, which develop extremely slowly. After birth, the young is carried and breast-fed for about 5 years. Weaning takes place only a few months before the next sibling arrives and then the youngster starts to build its own nest to sleep alone. They, nevertheless, remain with their mother. When they are about 10 years old, they will venture away from her for the first time. However, such adolescents on average still spend about 75% of their time with their mother. Females will leave their mother and community at about the same time, between 12 and 13 years of age, and start a new life in a different community. Adjusting to the new community takes some time, until the first baby arrives at around 14 years of age, if not older. Males leave their mother progressively when around 13 years old and are considered to be adult at about 15 years when they are accepted into the adult male dominance rank. As they start to leave their mother, they begin at the same time to dominate adult females and can invest a lot of time in this activity. Females produce an infant about once every 5 years and can continue to do so all their lives. Males rise in dominance with age, peaking at 30 years old when they slowly decrease in rank. Chimpanzees are considered old when they are about 40 years of age, by which time most of them are losing the hair on their heads and becoming old-looking in appearance. However, many individuals can remain very active and are still socially respectable members of their group.

What makes them chimpanzees? Like all animal species on this planet, chimpanzees have their uniqueness. These are the qualities we want to understand to get a grip on 'chimpanzeeness'. Tool manufacture and use, hunting for meat, cooperation and altruism between group members, competition for sex, food sharing, years-long learning of feeding techniques, individual identification in a social context, all these activities belong to the daily performance of chimpanzee life in the forest. Some of these abilities sound familiar to a modern human ear, others are more necessary for surviving in a forest environment, and a few are carried out by just one social group and not in neighbouring groups.

The more we learn about them, the more complex has become the answer to 'what makes them chimpanzees?', mainly because the more different populations of chimpanzees we study, the more diverse their behaviour has become. Compared with many animal species where one or only a couple of studies exists, the studies of chimpanzees are in abundance! However, from this luxury arise our doubts. A dozen chimpanzee communities are nowadays under study in different regions of Africa. The living conditions of these populations comprise some of the natural habitats typical for the species and represent some of the variability that exists. We will particularly concentrate on these in Chapter 6 to address the large social and behavioural differences that are documented. The important point here is to emphasize that, so far, there are not even two chimpanzee populations that possess the same spectrum of behavioural traits. Many more chimpanzee populations should be studied to understand the breadth of the behavioural diversity of the species.

In part, it is already too late! Chimpanzees are already extinct in two countries, while only small scattered populations remain in a few of the others. The chimpanzee's range comprises naturally dry savannah with limited gallery forest, as in southern Mali or in large regions of Tanzania, mixed woodland forest types, as in Gombe and Mahale National Parks in Tanzania and the wet tropical rainforest in Côte d'Ivoire, Gabon and Congo. Therefore, a dozen of chimpanzee communities under study is a depressingly small sample and represents only a small fraction of all the different ecological conditions they have faced during their existence in Africa. Many of the forests where chimpanzees used to live have been cut down and other populations have been hunted out. So chimpanzee populations have been reduced to a small fraction of what they were at the beginning of the eighteenth century. Furthermore, the continuing high rates of deforestation make the likelihood of obtaining a complete understanding of variability within chimpanzee populations an illusion. We are in the middle of a desperate race to try to know our nearest cousins before they disappear. Sadly, we are bound to underestimate chimpanzees' behavioural flexibility and have to make the best of the knowledge we have at present.

Furthermore, independent from any ecological differences, cultural differences in chimpanzees have now been documented in many different domains that increase behavioural differences in this species in a yet unknown dimension. For example, different behavioural patterns for a same task are observed in different populations of chimpanzees. While Falstaff hit the tick he found on Brutus on his forearm, a chimpanzee at Gombe would have placed it on a pile of leaves and squashed it with his fingernails, or a chimpanzee in Mahale would have placed it on a single leaf and folded the leaf in such a way as to cover the tick and then cut the leaf blade with his fingernail. Why these differences? What difference does it make to squash a tick on a forearm or a leaf? That is how culture is in chimpanzees – arbitrary, unpredictable and shared by group members![7] Furthermore, similar types of differences have been found between communities of chimpanzees within a same population. For example, while Salomé would dig her arms up to the shoulder in a driver ant nest to extract as many grubs as possible, Eva in the South Group just 4 kilometres to the south would enter just her forearm in the same species of nest and take a single little handful of grubs. Similarly, when young Kendo wants to attract the attention of a sexually active female he will knock a tree trunk with his knuckle, while young Gogol in the South Group will make a quick ground nest to attract her attention. We have searched for possible ecological factors that may influence this behaviour, but there is no explanation for the differences. Individuals within a social group conform to the group habit and seemingly for no other reason. Cultural differences, different ecological adaptations and varying complexity of social groups all influence the development of the individual chimpanzee in his or her natural group. To answer the basic question of 'what makes a chimpanzee?', we need not only a precise knowledge of chimpanzees within one group, but also an understanding of how variable their behaviour may be under different natural living conditions.[8]

OUR FORESTED PAST

Raymond Dart in the 1930s first proposed that our ancestors evolved towards human once they moved out of the forest and adapted to open savannah conditions. The challenge of adapting to this new environment was the main reason why our ancestors acquired typical human abilities, such as tool use, cooperation in hunting for meat, extended sharing between sexes and with infants, division of labour and warfare. Raymond Dart's theory gained support and was widely adopted. For a long time, it was supported by new findings of fossils of our oldest and more recent ancestors in the human line in East and South Africa.

It has since been realized, however, that this 'savannah' was often covered with trees! Over time, the climate changed and many of the areas that are savannah today were once covered by forest or at least woodland. For example, many of the places where the australopithecines and early *Homo* lived were forested and the associated fauna was typical of such a forested environment. More detailed and thorough analyses of pollen remains associated with these fossils have revealed that many were from forest plants and that the habitat at the time was much more heavily wooded than previously thought. Much of our past was spent in forest.

Chimpanzees still live throughout the tropical forest belt of Africa and the important population differences we observed are going to be the focus of Chapter 6. In my view, trying to understand, in a species very close to humans, how the environment affects some behaviour patterns that have until recently been thought as uniquely human, will be an instructive way to think about how ecology and evolution might have affected behaviour in the human line. However, before doing this, and after having detailed the life of the Taï chimpanzees in the forest, I'll expose in the next chapter some of the most dramatic aspects of the chimpanzee's life: intergroup competition with its multiple facets of violent behaviour, on one side, and, on the other side, the various extraordinary cooperative and altruistic sets of behaviour.

NOTES

1. The debate is still ongoing as to how large these divergences are between different chimpanzee populations (Fisher *et al.* 2006, Becquet *et al.* 2007). Does it justify a subspecies distinction? The problem lies mainly in the fact that different morphological or genetic traits have different degrees of divergence. For example, neutral genetic markers will diverge more than the ones strongly under the influence of selection. It is the same for morphological traits: for example, the morphology of jaws will tend to vary less than that of legs. Scientists still have not reached agreements, however, about how to value these different traits to make sensible taxonomic conclusions. So, depending on the traits or genetic markers one considers, conclusions differ. To me as a field worker who has seen many populations of chimpanzees in West, Central and East Africa, they clearly look very similar, both physically and behaviourally and the subspecies distinction seems not to reflect what is seen.

2. Fossils are by definition discovered where the conditions are appropriate for fossilization. So the existence of fossils does not give a fair representation of where our ancestors actually lived. We know that fossilization is only possible under specific conditions including rapid burial of the remains and an absence of humidity. So forest regions are by definition poor regions for finding fossils. As expected, most fossils of human ancestors have been found in dry regions and around lakes or near volcanoes where rapid burial of the remains could occur. However, it is suggestive that these fossil-rich areas surround the large central African forest regions, so it is possible that our ancestors lived in the forest regions. The few people that have excavated in forest regions were surprised to find older than expected human remains (Bailey *et al.* 1989, Mercader *et al.* 2002).

3. Giving a precise date for the separation of our ancestors from the chimpanzee line proved to be more difficult as most recent and complete studies have

shown that separation might not have happened in 'one go', but could have been an extremely long process. For example, a recent study shows that chimpanzees and humans presented the very first signs of separation less than 5.4 million years ago but with clear signs of interbreeding between the two lines still occurring up to some 1.4 million years ago (Patterson *et al.* 2006). Such a complex divergence process with initial divergence followed by genetic exchanges for a relatively long time has also been observed in gorillas (Thalmann *et al.* 2007).

4. The exact nature of this fission–fusion system in chimpanzees remains to be totally determined. In Taï chimpanzees, we had a good view of it and how it reacted to demographic changes (Boesch and Boesch-Achermann 2000, Lehmann and Boesch 2003, 2004, 2005). It is intriguing to see that in East Africa many populations of chimpanzees present a level of flexibility with much lower sociality in the females and possible spatial subclustering where some individuals have distinct ranges compared with Taï chimpanzees; for example, some females in Mahale and Goualougo chimpanzees are never seen in certain areas of the group territory (Nishida *et al.* 1990, C. Sanz and D. Morgan unpublished data), or in Gombe some females are very peripheral, possibly interacting regularly with two different communities (Williams *et al.* 2002).

5. The presence of male kin was often cited as one of the arguments for the development of cooperation in our human ancestors (Foley 1995); however, genetic data and modelling of this situation show that because all mothers transfer between groups this is not the case and males are not more related than females (Vigilant *et al.* 2001, Lukas *et al.* 2005).

6. Incest avoidance in social species was first thought to be a general phenomenon and achieved by the dispersal of the sexes at maturity. However, this mechanism is not totally successful as usually only one sex disperses and some incestuous mating and births have been observed in Gombe chimpanzees (Constable *et al.* 2001) and Karisoke mountain gorillas (Bradley *et al.* 2004).

7. Culture in chimpanzees is worth a whole book and is one of the most exciting developments in the field of psychology, anthropology and philosophy (Whiten *et al.* 1999, Boesch 2003). For purposes of this book, I limit myself to mention it here as one contributing factor to developing population differences in chimpanzees. I will in a forthcoming book specifically address the issue of culture in chimpanzees.

8. Population difference is a notion that is not always well recognized, especially in experimental sciences. Captive animals grow and live under very special and often unnatural conditions and, therefore, develop behavioural patterns and abilities that are not representative of their wild counterparts. Nevertheless, studies carried out with captive animals are too often taken as representative of the whole species without much questioning. Progress in some disciplines, like comparative psychology, runs the risk of been impaired by a lack of understanding of the role of the environment in the development of individuals (Boesch 2007).

5

Make war to get love

Warfare is a typical human characteristic that sets us aside from all living species because of its mixture of horrific extreme violence combined with heroic cooperation among warriors. Humans are a strange species, a species that through civilization has lost the gentle and peaceful nature of the noble savage, considered by some to characterize our ancestors – a nature we shared with our cousins, the great apes. Presumably like you, I believed this kind of statement to which most experts seemingly agree as it can be read in every textbook.

That is until the day in 1987, when I followed a large party, including Brutus and five males, in the hope that they would hunt. They rested under a group of red colobus monkeys in the far east of the territory. At half past two, they all jumped to their feet when they heard strangers drumming far off to the east, and responded with a few short screams and reassurance behaviour among themselves. They moved towards the drumming. They were quickly joined by a seventh male and eight females, most of them carrying babies. They formed a tight group and advanced silently and rapidly for 5 minutes in a straight line in the direction of the earlier drumming. Then, Brutus led them slightly away from the main group of strangers. All the females, with Kendo and Ulysse, now lagged behind, which is the typical behaviour of the females in such situations. They follow for a while as support but avoid the last, often physical, part of the attack.

Brutus with old Falstaff, Macho, powerful Schubert, Rousseau, one juvenile male and an adolescent immigrant female now progressed slowly in a line carefully listening for any signs of the strangers. They also looked regularly to the south-east from where a few calls of the strangers could be heard, presumably from the original main group, apparently unaware of their neighbours' approach. Fifteen minutes after the first signs of the strangers, Brutus, the leader, seemed to notice something ahead and started to run quickly and silently followed by the other six chimpanzees. I followed as best as I could trying not to make too much noise. About 100 metres ahead, the attackers found some of the strangers, as I heard violent barking and the anxious screams of two chimpanzees. The pursuit lasted for a minute, and eventually all the barking concentrated on one point. I joined them just as Kendo overtook me, moving at full speed and barking aggressively. A stranger female with a 3- to 4-year-old infant clinging to her belly had been trapped and was now surrounded by Brutus and his team of males, while the adolescent female drummed nearby. Holding her by a foot and one hand, they prevented the victim from making a move. She screamed and barked at the males. They hit and bit her on the head, shoulders and legs as she faced the ground, shielding the infant with her body. I did not see any blood on her or on the ground, but I was too far away to be sure. They did not torment her for more than a minute at a time, and let her rest for a moment in between, but pulled at her leg whenever she tried to move away. The hands and feet of the infant remained visible throughout the whole attack, but I never saw any of the males grab or bite the infant. About 5 minutes after she had been captured, three or four stranger males, aggressively barking and screaming, arrived to rescue her. Immediately, Brutus, Rousseau, Kendo and Snoopy faced them in a line. The female rushed towards her males. For some seconds, the two parties faced one another. Then, Brutus and his team attacked and the strangers disappeared without a sound.

Elaborate team work, leadership in attack, waves of coordinated attacks, the deliberate searching for enemies in a weak position, the taking of prisoners, and support by teams, systematic territorial defence, and patrols making deep incursions into foreign territory, etc. – all very war-like! Could it be that chimpanzees have been engaging in warfare for a very long time, longer indeed for it to have been a human innovation? It would not be the first time that something that has been claimed to be unique to humans has been usurped by the discovery of an earlier origin, as was the case, for example, with tool use and hunting, two sets of behaviour shared uniquely between chimpanzee and man. Warfare, however, is another much more complex story.

What is war? As so often, specialists do not totally agree on a definition but they all at least stress war as being a 'socially organized armed conflict between members of different territorial units'.[1] Some emphasize the killing itself, while many of the war activities of humans, such as patrolling, may not involve killing. Raymond Kelly proposes an evolutionary scenario in which he sees a transition from peaceful non-violent human societies to lethal armed conflicts, starting first with spontaneous armed conflicts over access to resources followed by revenge-based conflicts with capital punishment in which an armed group attacks to kill a murderer in another group. Finally, war, including feuds and raids, is grounded on the principle of strong group membership, so that the killing of one individual is perceived as an injury to the whole group and retaliation against any member of the other kin group is legitimate. So, in humans, the initial collective violence targeted at trespassers, poachers and murderers gives way to retaliatory violence against the compatriots of such malefactors. Chimpanzees certainly do not use weapons very often, but they carry their natural weapons with them, and as we will see, canine teeth can be extremely effective. As shown above, chimpanzees strongly defend their territories against trespassers and do so as a socially organized group. But do these intercommunity conflicts in chimpanzees effectively 'legitimize' the killing of strangers? In other words, do they consider members of other chimpanzee communities really different from their own group members?

First we should know how 'normally' chimpanzees behave during within-group conflicts. Group conflicts are abundant and result in numerous minor injuries that heal within days. Males, especially, fight over access to females as mentioned earlier and these fights are rarely more than bluff displays with charges resulting at most in a few slaps. Males also fight for dominance that will directly influence their reproductive success and these fights are much more serious. Dominant individuals are regularly greeted by lower ranking ones and the social life remains mostly peaceful. However, if low-ranking individuals want to challenge the dominant ones, fights are the established means of doing so. The most violent are fights for the alpha position and can be dramatic when the alpha position holder is still young. Macho was only 25 years old when Kendo started challenging him for the alpha position. The situation was tense for many months with many charging displays but rarely fights. Then one day, Kendo probably judged that the situation was changing in his favour and he viciously attacked Macho. In the fight, Kendo bit off a finger joint of Macho's right hand and a toe of his left foot. Macho was defeated and looked to be in a bad shape for a week, although it was not clear whether this was because he was depressed or hurt or both. These were the worst injuries I saw in all my years with the chimpanzees. Severe injuries resulting from fights have also been seen in Gombe and Mahale chimpanzees where wounds to the abdomen or on the back have forced individuals away from the groups for months, but lethal fights within communities are very rare.[2] Are the levels of violence different in conflicts between communities? What are the situations and the differences that result in increased levels of violence? Are strangers really different from insiders?

On 1 March 2005, Emmanuelle Normand, a student working on the Taï project, followed a party of chimpanzees from the South Group, including four adult males, two adolescent males and four adult females. Suddenly, they heard their East neighbours and after some reassurance gestures between them, they all ran silently towards them. When Emmanuelle caught up, the whole South Group, including three females, Sumatra, Zora and Wapi, was gathered around a young adult male chimpanzee of the East Group. They all bit and struck him violently. At one point, Sumatra was ejected from the team and hit by a male who jumped with both feet on to her back. After this, both of them returned to the attacking gang, screaming at the top of their voices. Twenty-two minutes after the beginning of the attack, drumming was heard from farther to the east. The whole South Group rushed towards it, apparently in a highly aroused attacking mood, leaving their tortured victim behind. He just sat, hardly able to move, and 2 minutes later Zora came back and sat next to the victim. He was covered in cuts around the mouth and ears, but his throat was still intact. Five minutes later, three of the four adult males, including Sagu and Gogol, returned as well and renewed their vicious attacks for a minute. The observer had the impression that this was the moment when the victim was fatally injured with wounds to the throat. He was now clearly unable to move and the whole of the South Group moved away. Gogol came back twice, followed by Sumatra who tried to chase him away from the victim, but he managed to bite the victim's arm. Emmanuelle heard the sound of breaking bones. Thirty minutes after the start of this attack, the young male seemed dead and his lifeless body was left behind, except for Sagu who returned for a last time 7 minutes later and struck the corpse.

It came as a shock to me and probably to most chimpanzee observers, that Gogol, with a single bite, could break the forearm bones of an adult chimpanzee. Even after studying them for all those years, I had not really been aware of the strength of a chimpanzee's jaw. Gogol was the beta male of the South Group, a young male not yet in his prime. As a young adolescent, he had developed a great infatuation with Jacobo, Julia's baby, regularly stealing him away from his mother and moving unconcernedly in the group with her baby on his back. Jacobo looked relaxed and happy to be carried around playfully by this large male, while the understandably anxious mother searched for him. Gogol fought his way up the social ladder through many fights. He was one of the keenest and best hunters in the community and also extremely generous, sharing meat with the many orphans in his group. Gogol's unrestrained strength against the stranger hints at how controlled the males usually are in their fierce fights within the group. This difference certainly indicates that within-group fights are altogether different from attacks against strangers.

Immediate inspection of the dead victim's body by Emmanuelle confirmed his death was probably caused by the large 10-centimetre gash inflicted by the attackers on the throat and upper chest. Furthermore, the testes and penis had been removed leaving a clear hole. Unfortunately, her view of the attack had been blocked by the many participating chimpanzees so she had not seen when and how the emasculation had occurred.

Violence with extraordinary force, killing of a male captive, mutilation and emasculation of the victim, coordinated group attack – this continues to sound disturbingly war-like! Something in the last powerful bite of Gogol, which broke the stranger's arm, is reminiscent of the so-called 'depersonalization of enemies'. This certainly throws a totally different light on the controlled aggression observed daily between chimpanzees in the same social group. The killing was quick and performed without hesitation although interrupted by calls from the strangers. The active participation of the females was intriguing as war is usually, although not exclusively, a male affair. Even more intriguing is the emasculation, the removal of the victim's penis and testes. This seems an unambiguous message about the purpose of such warfare: annihilation of a competitor for sex.

Brutus, Falstaff and his warriors have for over 15 years patrolled the borders of their territory and responded to numerous intrusions of strangers as an organized team, sometimes bewilderingly fearless, to defend their home or attack neighbours. I have been uniquely privileged in being able to follow them in hundreds of such expeditions but have never witnessed such killings. Brutus and Co. always showed an intense concentration in trying to find and defeat intruders and demonstrated, to my eyes, always a baffling readiness to face them under many different circumstances. Never did I see them trap a single male, and when they trapped a female, as we will see, they did not show unrestrained violence. Having read the descriptions of fatal violence from the Gombe chimpanzees, I started to wonder if there was not something different in the level of violence in Taï chimpanzees. In our monograph about the Taï chimpanzees published in 2000, Hedwige and I wondered about possible explanations for this difference.[3] So I was very intrigued and thrilled when early in 2005 Emmanuelle Normand told me about her observations of the killing of an East Group male by the South Group chimpanzees. Already aware that the chimpanzees of Gombe and Ngogo have been killing neighbours, what intrigued me most was that the ambivalence between violence and cooperation I have always seen in male chimpanzees was reaching higher levels of complexity than I had suspected.

Some of the detailed descriptions of deadly violence in Gombe chimpanzees were also suggestive of a difference in attitude towards strangers compared with group members. Never did Jane Goodall or other observers describe chimpanzees behaving towards group members in the same way as they did against strangers, where they were seen to bite a stranger in the face and drink the blood running down the face, or when they were breaking the articulations of a victim's legs by twisting them round and round. Gruesome descriptions of the dark side of the chimpanzees made their way into the popular literature such as in the bestseller *Brazzaville Beach* by William Boyd, but at the same time they revealed the highly complex social life of this species. In chimpanzees, aggression within the group can be very high but it simply never attains the same degree as when facing strangers.

On 4 March 2007, the second observation of intercommunity killing in Taï provided some further insight on aspects of warfare in chimpanzees. Camille Bolé, a year-long experienced field assistant of the Taï project, witnessed the adult male Porthos, a member of the East Group carrying Gia on his back, rescue the adult female Bamou from an attack by South Group chimpanzees (see Chapter 3). Three minutes after the South Group members had run away from Porthos, Camille heard another tremendous outburst of calls towards the west, while to the east the chimpanzees of the East Group still drummed and barked. Camille decided to head towards the calls from the west. A few minutes later, he arrived to see a mass of chimpanzees gathered around what looked like a large, dark animal. He then recognized Sagu, the alpha male of the South Group, who was looking at him with his mouth and teeth covered in blood. He realized that a large party from the South Group had captured another individual. All the adult males (Sagu, Gogol, Kaos and Zyon), four of the adult females (Zora, Isha, Wapi and

Olivia), and three adolescent males (Utan, Woodstock and Kuba) were present. All, except Kaos and Kuba, were hitting the captive individual and, except for Zyon and Utan, all were repeatedly biting the victim. He tried to fend off his aggressors, which only led to an increase in violence. Sagu, Gogol and Zora had blood-stained mouths from biting deep into the victim's flesh, while the other three females also bit the victim repeatedly but their mouths were not blood-stained. Camille then saw Sagu biting the throat of the victim for prolonged periods of time, which appeared to lead to the victim's death.

Fifteen minutes after the start of the attack, the attackers withdrew a few metres from the victim and Camille saw deep cuts on the face, throat and both sides of the thorax. The penis and testes had again, as in the description above, been neatly severed. The dead victim lay motionless on his back and the attackers looked around, while the East Group chimpanzees still drummed and called from the same position as before. The attackers responded but remained resting some 5 metres from the corpse. After 29 minutes, Gogol and Zora went back to the corpse and hit its belly while calling. The East Group replied. After 32 minutes, all the attackers rested 20 metres away from the body, Kaos and Zyon on their backs, while Wapi, Isha and Olivia sat facing the corpse.

A few minutes later, Camille noted that the female Zora was pulling on what he recognized to be the victim's penis and attached to it were the testes. Due to the extreme excitement during the attack and the constant changes in position of the attackers around the victim, he had not seen who had severed the genitals. Instinctively, we would assume that males remove the genitals from competitors and Zora could have received or stolen them from one of them. However, Camille had not seen such a transfer. Zora, with a blood-stained mouth, had been in the middle of the attack all the time and she might have severed the genitals herself. In any case, she ate them with fresh leaves taken from nearby saplings. He saw no one begging for any part and she ate the entire penis and both of the testes. Some of the chimpanzees now started to leave the place of the attack, but Sagu, Gogol, Woodstock and Olivia struck and bit the body again for 4 minutes. Gogol then remained with the victim's body. He bent the arms and legs at the joints and after extending them, bit forcefully into the insides of the elbows and knees leaving visible teeth marks. Thereafter, he bit all the phalanges of the fingers and toes, one after the other. About one hour and a half after the start of the attack, Sagu and Gogol pant-hooted, drummed and then left the site.

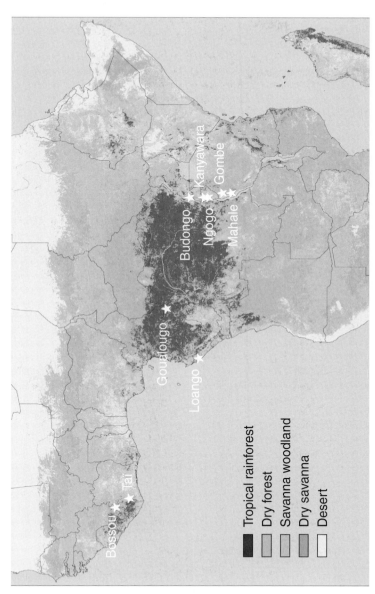

Plate 1   Vegetation map of Africa showing the main studied chimpanzee populations.

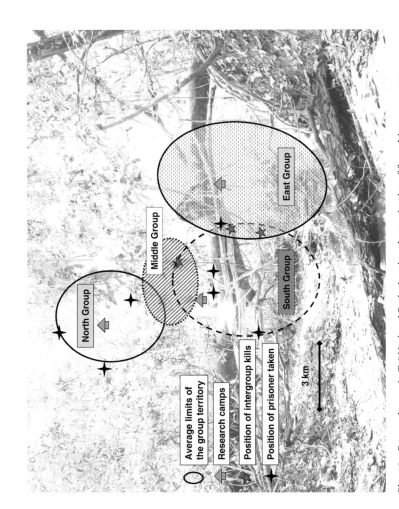

Plate 2   Research area in Taï National Park showing the territories of four chimpanzee groups.

Plate 3   Portraits of chimpanzees of the North Group in the Taï forest. (Photos by S. Metzger)

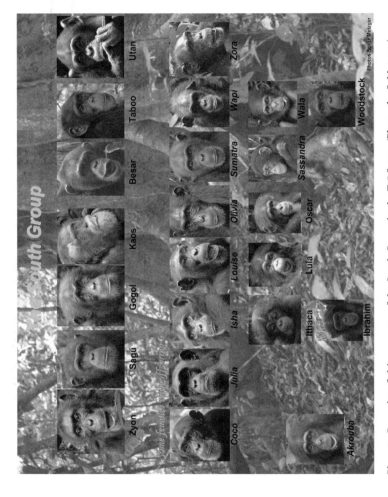

Plate 4 Portraits of chimpanzees of the South Group in the Taï forest. (Photos by S. Metzger)

Plate 5   Males of all ages are interested in sexually active females. Here the alpha male as well as a 3-year-old male infant are grooming a fully swollen female.

Plate 6   Leopards are an important threat to chimpanzees in the Taï
forest. Rousseau, an adult male, had just escaped a leopard attack that cut
open with its claw the scrotum skin of his right testis. The injury never
healed and Rousseau died two years after the attack, most likely from an
abdominal infection.

Plate 7 Brutus, in the middle, is being groomed by two of his female friends, Loukoum and Mystère, while he tolerates the young ones, attracted by the large male, playing on him.

Plate 8   Loukoum is using a wooden club to crack the hard *Coula* nuts on a large root used as an anvil, while 1-year-old Lefkas is watching her.

Plate 9   Once the nut is open, Lefkas directly begs with his hand
at the mouth of his mother to obtain some pieces of the food.

Plate 10   Three-year-old Papot is begging with success from his efficient mother, Perla, who cracked two *Coula* nuts per minute for over 1 hour per day during the 3 months of the nut-cracking season.

Plate 11   Lefkas, now 4 years old, rests near his sleeping mother. Dominant females such as Loukoum can invest up to 6 years in sons before reproducing again.

Plate 12   Fitz, the second son of Ella, developed with her support into a very strong and confident male. He allied with her in supporting his oldest brother Kendo and thereafter became the dominant male when only 15 years old.

Plate 13   The rainforest is full of insects that can sometimes be accessed only with the help of tools. Here two chimpanzees are using small twigs that they fashioned themselves to extract beetle grubs hidden under the bark in the wood of a dead fallen tree.

Plate 14   Marius, a young adolescent male, is eating some fruits while looking intently at a group of monkeys and evaluating the likelihood of making a successful hunt.

Plate 15   A capture of an adult monkey elicits high excitement in all chimpanzees present and is followed by a wild rugged scrum, in which all males present and some females want to get access to the meat.

Plate 16   Taï chimpanzees regularly hunt red colobus monkeys that are one of the largest species of arboreal monkey in the African forest. An adult red colobus monkey has just been killed by a group of male chimpanzees.

Plate 17   Brutus is eating an adult red colobus monkey and generously sharing the meat with Kendo, on the left, who contributed to the hunt, while Loukoum is looking at them.

Plate 18 The young immigrant female Fossey is tolerated by Brutus, in the background, to obtain meat scraps, although she did not contribute to the hunt. Females obtain more meat than non-hunter males, despite the fact that they hunt very rarely; it might well be a way for a male to increase a female's willingness to mate with him in the future.

Inspection of the dead body by Camille Bolé and pictures taken at a later time confirmed the victim as Aramis, a young male of the East Group. Aramis had been killed within 14 minutes. Some females of the South Group had taken an extremely active role in supporting their males and had themselves inflicted injuries on the stranger. Zora's consumption of his testes and penis left the observers perplexed and pondering about female chimpanzees and mythical Amazon warriors.

Fatal violence between groups or between owners of territories has been documented in many animal species and in this sense violence is nothing special. In most territorial species, the violence occurs between the territory holder and any intruders, and it stops as long as the territory limits are respected. Harem holders in gorillas and langur monkeys also fight very hard to protect their females and instances of deadly fights have been reported in both species, as losing a harem means the end of the owner's reproductive life. Group violence is rarer as it requires both group living and essential mutual interest. One of the most famous examples of such group violence is seen in the lions, where groups of adult males, often brothers, attack males controlling a pride of lionesses. The only way for male lions to reproduce is by controlling a pride long enough for them to produce offspring with the females and have the cubs grown up safely. Pride owners are normally not able to acquire another one once they lose one, and therefore they fight very hard to keep theirs, so that many pride take-overs result in bad injuries that can be fatal to the previous pride owners. Interestingly the need to be a large, and therefore successful, coalition to take over prides makes brothers associate and fight together. The unlucky ones that have not enough brothers would associate with unrelated ones but such coalitions are smaller and less successful.[4] In large lion coalitions, reproduction is largely favouring one individual at the cost of the other brothers. In contrast, in chimpanzees, group violence involves many more adult males, who are rarely brothers, and the large coalitions they form use many different strategies to confront, over many years, up to five to six different neighbouring groups. In addition, the benefits of such confrontations are still a matter of debate as females and food resources do not directly change ownership following such fights, as is seen in lions.

Having followed the chimpanzees in the forest for so many of such intergroup fights, I was always impressed by the level of organization and solidarity between the males during such highly exciting and risky actions. For over 15 years, I have been with Brutus, Falstaff and Kendo, the warriors of the North Group, and have seen how they planned such aggressive contacts and made very bold incursions. If they were tense and intense, they never gave the impression of being hotheads who did not consider alternatives and weigh the odds carefully. Having read about chimpanzee violence by Jane Goodall describing the warfare of the Gombe chimpanzees in her book *Through a Window* (1995) and in the *National Geographic* magazine, I was not caught by surprise but puzzled as to why chimpanzees resort under some circumstances to such a level of violence and why they seem to use it with different degrees of cooperation. Could it be that chimpanzees strategically plan their violence and use it to achieve precise goals? Could it be that chimpanzee warriors are more like strategists weighing the pros and cons of confrontations?

HOW DO CHIMPANZEES MAKE WAR?

In all simple hunter-gatherer and forager societies, the predominant war activity is the raid in which a group of men tries to surprise and harass their opponents by dealing them a deadly blow and then retreating as quickly as possible to avoid any casualties. There is a strong preference for taking the least possible risk in warfare and, therefore, raids are aimed predominantly at weaker, defenceless individuals. To achieve this, opponents are surprised and ambushed and can be attacked when unarmed, such as when they are asleep. This has been abundantly documented in the Aborigines of Australia, in different societies in New Zealand and Tasmania, as well as in South America (Yanomamö, Ache, Jivaros, etc.), the North American Indians, the Inuits and many different societies in Africa (Nuer, Hadza, etc.).[5]

Patrolling and raiding certainly accurately describe the main strategy used by chimpanzees in the Taï forest, as they defend their territory from neighbouring stranger groups. Before I detail some of these strategies, some background information about the territorial behaviour of Taï chimpanzees is necessary. The Taï forest is a continuous stretch of rainforest of over 4,500 square kilometres, with numerous small rivulets all running south. The only interruptions to the tree canopy are caused by the many natural treefalls, encountered on average every 50 metres. Thus, the limits of chimpanzee territories are determined purely by the strength of the different chimpanzee communities and not by the topography of the forest. Larger communities have larger territories, as more males are able to patrol and defend them. The territories do not have strict borders as there is a relatively large overlap between neighbouring communities. Unlike chimpanzees, a large number of territorial species maintain strict boundaries and no trespassing is allowed. However, the larger the territory, the more difficult it is to defend its boundaries.

Patrolling is the only way by which a 20 square kilometre territory can be preserved from incursions by neighbours. Therefore, chimpanzees of the Taï community were seen to be involved in territorial activity at least twice per month all year round. A patrol could start at any moment and often was initiated without any signs of strangers, but simply because the chimpanzees were already near the periphery of their territory. In about a third of the territorial activities, the males actively went on patrols to search for strangers and, in most of the cases, these patrols took them deep into the territory of their neighbours.[6] As seen in Chapter 3, they can develop over some hours. I had the impression that a real effort was made to locate strangers, even if they only succeeded in a quarter of the patrols. Typically, when a group was foraging near the edge of their territory, the males left the females without making any particular calls and started a patrol. The adult males were the main patrollers and their disappearance was often sudden and silent. They may wait at a distance for more males to join them for, as a rule, fewer than four males did not go on patrol (96% of the patrols had four or more males). In other situations, a whole party with females and babies may start to make a deep and silent incursion into the neighbouring territory.

A patrol is typified by the resolute and silent progression of a tight group of male chimpanzees: they advance rapidly in a line, one behind the other and stop regularly to listen and search for signs of chimpanzees. One day, on 29 September 1993, I was with a large group of females with infants and some males. After having rested for a long time on a large windfall tree in the southern part of their territory, they started to move towards a ridge. Keeping an eye on the males, I noticed some exchanges of glances and suddenly, at 10:00 hours, Macho, Kendo, Fitz and Darwin moved silently towards the east (at this time there were only five adult males alive in the community). They moved decisively, crossing a valley and following the ridge on the other side. They listened regularly, sometimes sitting for about 3 minutes. Then they entered a valley, 2 kilometres inside the strangers' territory. Forty-five minutes later, they showed some interest in black-and-white colobus monkeys and made two silent hunting attempts. They then turned north.

Two hours after the start of the patrol, they all rested silently within a windfall. Suddenly 10 minutes later, the strangers screamed not too far north in the valley. Immediately, the four males stood upright with bared teeth and reassuring one another. Then they all moved towards the strangers, but after only 100 metres they stopped to listen, to make sure that they were going to surprise the strangers and not the other way round. Since the start of this patrol in the morning, Darwin had led the party. After 10 minutes of careful approach, they started to look at where they were placing their feet and hands, so avoiding making any noise. I did my best. Twenty minutes later, they spotted the strangers in a tree and continued to move forward. I saw the strangers feeding in a fig tree, with some infants playing in the branches, and all of them totally unaware of the presence of the other chimpanzees. Fitz sat and looked at them, partly hidden by the foliage. Darwin and Kendo positioned themselves on either side of Fitz, while Macho remained a little way behind them. Obviously they were waiting for the strangers to come to the ground, but the strangers were resting now and we heard their lip-smacking as they groomed one another.

Suddenly, 20 minutes after looking at the strangers eating and resting in the tree, Darwin got to his feet and silently threatened the strangers by waving his arm. Immediately, Fitz, Macho and Kendo barked and rushed to drum on the fig tree. Kendo climbed the adjacent tree to threaten them at closer range. Somewhat surprised, the strangers remained at first totally silent. After one minute, they suddenly all barked and screamed aggressively. Kendo rushed down, but the strangers did not move and the four males came together again under the fig tree and drummed. Once more the strangers called aggressively without moving. Three minutes later, an adult male rushed to the ground and ran away. Kendo and Macho immediately chased after him and, judging by the intensity of calls, they seemed to catch up with him. Hearing that, the strangers called and all rushed to the ground in a close pack. I counted five mothers, two adult males and three more chimpanzees, all probably going to rescue the first male and pursued by Darwin. Fitz remained under the tree looking up and threatening those that remained. From what I could hear, the fight in the north had changed into a pursuit. Now an infant screamed tenaciously in the fig tree as he did not dare to follow his mother and felt trapped. As a response and in support, three stranger males counter-attacked, which allowed the infant to rush down with another adult male who had also remained in the tree. For 3 minutes, the two groups of four males chased one another, with the two brothers, Kendo and Fitz, leading the attacks. Soon, the four males of the study community headed back westwards, drumming repeatedly and loudly.

The strangers moved east and after 10 minutes, they were no longer heard, while the four drummed for another 5 minutes. One hour later, they joined the initial group of the morning.

The further a group intrudes into the neighbours' territory, the more carefully it progresses and the more regularly it stops to listen. In this forest, sight does not help to locate strangers, but listening is very important. Patrollers remain silent during the whole time, occasionally eating a leaf or two. At a crucial point, such as a ridge, they may stop and listen for some time before going on. They cover large areas within the stranger territory and normally return to their territory at a different point. They regularly sniff at tree trunks and leaves, more so if they come across fresh traces of chimpanzees, such as a chewed wadge of fruit, a nut-cracking atelier or a nest. Some may climb up to a fresh nest but leave it undisturbed. The end of a patrol, when re-entering their territory, is usually signalled by very loud drumming. The only feeding activity that we saw in a quarter of the patrols was hunting for colobus monkeys. In stark contrast to hunting within their own territory, they remain silent throughout the hunting episode and if a squabble breaks out between them, the screams that normally are so loud are totally suppressed.

Besides, patrols are a risky activity for chimpanzees, for there is no additional help to count on if they become outnumbered by strangers. The size of a party in Taï chimpanzee is on average 10 individuals, whereas two thirds of patrolling parties include fewer than five individuals, mostly only males, so being outnumbered is quite likely. It seems that the goal of patrols is not to win battles against the neighbours, but rather to gain information about their location and, if they find them, to try to unsettle them. As just seen, they would make a surprise attack, and pursue the strangers, but when counter-attacked, they simply run away without putting up any resistance, until they are no longer pursued and return to their territory.

What happens when they suddenly detect the presence of strangers? Or when they hear the calls of strangers? On 1 July 1985, 25 chimpanzees, including all the adult males of the North Group, moved as a tight group north-west in the northern part of the territory, possibly after hearing strangers. Ten minutes later, I heard the strangers drum not too far away to the north. Without a sound, Brutus headed directly towards the strangers. Five minutes later, one of the chimpanzees shrieked with fear. Brutus and others immediately quietened him. At 13:10 hours, the strangers called from the north-east and they all turned silently in this direction, walking carefully to avoid twigs that might crack under their feet. Twenty-five minutes later, we heard the calls and screams of normal social activity among the strangers. Brutus reassured Schubert, the beta male, and they moved on. The adult females regularly sought reassurance from the adult males. All scrutinized the vicinity carefully for any strangers. We could now hear them feeding on leaves in a tree. Four of the ten adult females turned back and headed southwards. The males, from a distance of about 40 metres from the strangers, fanned out in a line with Brutus at the extreme left and Schubert on the right. More females now headed back. Schubert seemed to hesitate, while Brutus, with very loud aggressive calls, started the attack, instantly followed by the males beside him. Schubert followed, but some metres behind. The last females now left, as the males started to attack. There were many strangers and the attackers screamed in surprise as the strangers climbed down from the trees. Within a minute, the attackers were chased away by four of the stranger males. A second wave of strangers followed but they saw me and made a detour before running to support the males in the front of the attack. In the second wave, I saw two stranger mothers with infants on their backs. The attackers faced the strangers about 100 metres away; they counter-attacked briefly, but the strangers pursued them again over some 200 metres, and the attackers retreated rapidly. After hearing the strangers, the chimpanzees always reacted, but the reaction was a function of the number of individuals within the community: as long as the North Group had more than six adult males, they attacked the strangers in two thirds of the cases. This proportion decreased to less than a quarter when fewer than six males were present, and if fewer than three males they would retreat after hearing strangers in two thirds of the cases.[7]

Whenever the forces in presence seem to be fairly equal, back-and-forth chases between opponent groups occur quite regularly, so as to form front lines and make repeated attacks followed by counter-attacks. One day in 1986 in the far east of their territory, the males of the North Group heard strangers and, as a large party, silently went to attack them. Getting close, they saw the strangers and Brutus initiated an attack with loud aggressive barks, followed by all the others. They disappeared from sight, but then I heard wild calls and saw the North Group males running wildly towards me pursued by about seven or eight stranger males. Brutus and the other chimpanzees stopped where I was standing and the strangers also came to a halt, barking loudly. The males formed two parallel lines about 15 metres apart. Brutus then charged again followed by the other males. The strangers ran from sight pursued by Brutus and Co. Louder calls were heard and the process was repeated. The third time, when the North Group line was now 10 metres in front of me, they were joined by two young females in full oestrus, Bijou and Mystère, and a mother, Tosca, with a baby. While both male lines remained facing one another, the three females walked quietly to the other side, and just behind the line of the strangers, mated with some of the stranger males. At the same time, I saw one young female of the stranger group do the same, joining the North Group males and mating with Macho. After some minutes, the two groups parted, and I followed the North Group back into the heart of their territory where they did some loud drumming.

I have observed a few such male front lines making back-and-forth attacks that gradually, once the overt aggression had quietened down, changed into displays with some females mixing with the males. However, normally the attacks are too quick and visibility too limited to permit a precise evaluation of the forces present.

Surprising chimpanzees while they are up in trees happens regularly and can have fatal consequences. On 8 September 2002, at 16:06 hours, Camille Bolé with his colleague Nicaise Oulaï followed a large party of five adult males and nine adult females of the South Group, as they moved north into the western part of the territory of the Middle Group, when they heard the neighbours drum to the west. They advanced silently in the direction of the drumming, listening carefully and looking in all directions. After 6 minutes, two South Group males drummed and pant-hooted and, in the west, the neighbours replied. The South Group advanced, drumming twice more. After 21 minutes, they surprised a party of the neighbours in a fruiting tree. They all fled westward pursued by some of the South Group males. A young male, Sagu, spotted a whimpering infant in the tree that had been abandoned in the panic. He climbed the tree, captured the male infant, pulled him down with him and dashed the screaming infant to the ground. The neighbours made loud calls in reply to the distressed screams, but the only adult female who came into view was immediately chased away by some of the South Group chimpanzees.

Sagu climbed the tree again and hit the infant against the branches, watched by some of the other chimpanzees from the ground. Ten minutes later, Sagu started to bite the infant and the sound of breaking bones could be heard. Later he twisted a foot of the screaming infant. In response, the two groups exchanged calls, and Sagu let the agonised infant fall to the ground. Seven adults of the South Group watched it carefully for some 12 minutes. Another young male, Gogol, threw him around, but a female, Duna, groomed him. Gogol also groomed the infant for a little while, after screaming from the neighbours was heard. One minute later, Gogol bit the infant in the leg but, attacked by Duna, he ran away. A quarter of an hour after its capture, Gogol bit the infant in the throat, most likely killing it, and then he broke its fingers, feet and some of the joints. He extracted some bones from the feet but did not eat them. After 11 minutes, Gogol followed the others leaving the corpse behind, but Duna came back 2 minutes later and dragged the corpse with her as she joined the others feeding on fruits of the *Sacoglottis gabonensis* tree. The neighbours called again in the west. One minute later, the last chimpanzees of the South Group moved towards the south, returning to their territory with some drumming, leaving the dead infant behind on the ground.

Infanticide is a common occurrence in many animal species including humans. The classic example is the killing of all infants when male lions take over a pride of lionesses. The new males kill all suckling infants to force the females to come into oestrus for the males to sire them. In this way, they do not invest any effort in the infants of the previous males, thereby increasing the number of their own infants. This certainly is not a benefit to the females but their ability to respond to such powerful males is extremely limited. In primates, langur males do the same when they take over a new group. Gorilla males also try to kill the infants of females of neighbouring groups and sometimes these mothers will leave the group where the infanticide could not be prevented and, more rarely, have been seen to actually join the males that had killed their infants. In lions, langurs and gorillas, infanticide seems to provide some benefit to the infanticidal males.

However, in the case of the chimpanzee, the situation is less clear. If the killing of adult stranger males together with emasculation reveals something of their motivation, the reasons for killing the infants of stranger females is less obvious. We have no confirmation that a female would be more inclined to mate with or transfer into a group of the killers of her baby. In some chimpanzee populations, the killing of infants of stranger females has been observed several times[8] but, as is the case in Taï, no obvious reason for these killings has been found. The fact that most observations were done with non-habituated females made it simply impossible to track the reaction they showed subsequently to such a dramatic event. In Taï, the fact that the infant had been left unattended might somehow explain the subsequent killing. It is intriguing, however, that when female Taï chimpanzees with infants were captured, the attackers did not harm the infants.

Revealing of their motivations, when it comes to intergroup interactions, Taï chimpanzees were regularly seen to make stranger females prisoners. Observations are often fragmentary due to the low visibility in the forest and our attempt not to affect by our presence the outcome of the interactions.[9] However, when interactions between two habituated groups occurred, undisturbed detailed observations were possible. This was the case on 30 April 2000, when Catherine Crockford followed the three adult males of the Middle Group, Urs, Bob and Léo, moving quietly north into the North Group range. Suddenly at 16:21 hours, they rushed forwards and surprised two adult females of the North Group, Goma and Fossey, with their two infants and two juveniles. Goma, whose ability to run had been impaired for months by a handicap of both hips, bent over to shield her 4-year-old daughter, Gisèle, while the males displayed around her, hitting and jumping on her for a minute. Fossey and a juvenile disappeared immediately, while Fossey's 6-year-old daughter remained sitting and observing from a branch in a tree. The aggressive actions of the males towards Goma stopped quickly. Then, Urs made his first friendly contact with an open-mouth kiss, followed by Bob 2 minutes later. All three males performed genital inspection on Goma who had no sexual swelling and was lactating. Four minutes after the initial encounter, Gisèle left her mother to greet Léo, the alpha male of the Middle Group. Gisèle tried repeatedly to greet the males, but Goma whimpered and tried to make her come back to her. Urs tried without success to mate with Goma. After 15 minutes, Fossey's daughter also disappeared, leaving Goma and Gisèle alone with the males. When Goma tried to climb a tree, Urs struck her many times but then open-mouth kissed and groomed her. Léo successfully mated with her after 45 minutes. The males fed on *Treculia* fruits for a while, but Goma just watched them. When it was totally dark at 19:30 hours, they all made their nests in nearby trees, the males making theirs close to Goma's nest.

The next morning, the males tried to lead Goma back southwards into their territory but it took Goma 2 hours to leave her nest and get down to the ground. Urs displayed twice at her, and, for the first time, Léo hit Gisèle. Afterwards many reassurance gestures were observed between Goma and Léo, as the males tried to lead Goma eastwards into their territory. Léo mated twice with Goma, while Urs tried twice more without success. Three hours after leaving the nest, Goma tried to move back westwards towards the North Group territory. Léo mated with her and Urs and Léo groomed her. Gisèle and Goma received remains of fruit from Bob and Léo and ate for the first time since their capture. At 12:22 hours, for the first time, a drumming from the North Group was heard far to the west. The males tried to lead Goma to the north-east, whilst Goma tried to head in the direction of the drumming. As a result, little progress was made in any direction. Around 14:30 hours, Goma started to move west for short distances, but the males twice caught her. On the third attempt, at 14:45 hours, the males reacted too slowly and Goma escaped towards the west and hid in a dense treefall. The males did not try hard to find her. Eighteen minutes later, Marius, the alpha male of the North Group, drummed close to the Middle Group males. The three Middle Group males retreated southwards back into their territory, eventually drumming and pant-hooting.

Female prisoners were in the vast majority seen to mate with their male aggressors even when, like Goma, they were not having a sexual swelling. Instructive observations concerning the motivation of the males! We observed eighteen such kidnappings of adult females by males from neighbouring groups. Normally the captives were not held for long, as in the case of Goma, for the noisy support of the victim's group members arrived within minutes. Since poor visibility in the Taï forest prevents an accurate assessment of the rescuing forces, the kidnappers always faced rescuers at first, thus giving the victims time to escape. Interestingly, in none of the instances of kidnapping were the infants of these females directly threatened or hurt by members of the attacking community.

On 2 May 2000, the day after Goma escaped from her Middle Group captors, the entire Middle Group with three adult males and three adult females slowly and silently returned to the same area in which they had surprised Goma. As they approached, they heard the drumming of North Group members in the north. They listened silently, advanced for 8 minutes and then stopped and listened for 10 minutes. Remaining silent, they headed south back towards their own territory. Three days later, on 5 May 2000, all three Middle Group males initiated a patrol of the North Group territory, starting at the same *Treculia* tree where they had captured Goma. They patrolled deep into the North Group territory, right into the heart of the area, drumming once. They remained there silently for 2 hours, listening attentively. They drummed once again as they left the area. No reaction was discernible from the North Group at any time. Six hours after the start of the patrol, they returned to the border area of the two territories, to the same *Treculia* tree.

On 7 May, Urs and Bob from the Middle Group returned again to the *Treculia* tree. They approached the tree slowly and silently and stopped to sniff at some leaves. Both appeared unaware that Marius and Nino, the two adult males of the North Group, were silently sitting some 25 metres to the north, watching them intently. After 2 minutes, Urs on looking around made an initial movement to retreat at which point Marius and Nino immediately charged them. Urs and Bob fled southwards back into their territory with Marius and Nino giving chase and barking loudly for about 2 minutes. A minute after losing sight of the Middle Group, Marius drummed. Five minutes later, Urs and Bob drummed after retreating several hundred metres to the south. Both groups of males continued drumming for some time. Next, Marius and Nino came back to the same area and rested. Two hours and 50 minutes later that same day, the third and dominant male of the Middle Group, Léo, moved alone northwards, whimpering and pant-hooting, seemingly looking for Urs and Bob. He arrived in the same area as the morning encounter and ate two *Sacoglottis* fruits. Suddenly, he stood upright with his hair on end, while looking around intently. After 2 minutes, Marius and Nino appeared from the undergrowth in a full charge towards him. He screamed loudly and fled south back into his territory. Marius and Nino chased him for 40 metres then stopped to bark and drum. Léo continued running southwards, screaming continuously for 4 minutes and Marius and Nino continued drumming for an hour.

Following the capture of Goma for over a period of 7 days, the North and Middle Group males continued a tug-of-war contest testing their respective forces. The behaviour of Marius and Nino is intriguing. They were not present when Goma had been captured but nevertheless they waited for the kidnappers to return 6 days later to the exact location of her capture. How did they know that Goma had been kidnapped there? Did they anticipate during those 6 days that the kidnappers would come back? Did they want to take revenge for what happened to Goma? As so often, we are not able to answer all the questions, but what seems important here is that at least some of the intergroup interactions did not happen independently of each another and that there seemed to be some sequential development. It is not uncommon for chimpanzees to return to the place where they had had an encounter 2 or 3 days previously, as if to check on or show a wish to continue with the earlier event or to try to settle the issue. In humans, such war-like encounters are classified as 'revenge' and the original offence can be lost in the mists of time. I am not suggesting that this is what happens in chimpanzees but rather that the situation is reminiscent of human behaviour in the sense that one single confrontation does not resolve the issue.

In the Taï chimpanzees, prisoners were taken regularly, and in 11 of 13 cases sexual activities were observed between the female and her male aggressors. There is no doubt that the female prisoners were treated roughly; they were hit and bitten, but we had the impression that there was never any intention to harm or kill them. Most of the blows were concentrated on the head, hands and genitals but rapidly ceased if the female remained quiet and did not try to escape. When we did find some blood at the site of an attack, the injuries must have been superficial since none of the females was ever impaired in her ability to move. Attacking males did not give the impression of wanting to harm the females as whenever the victims were with their infants, the little ones were never directly threatened or hurt. In a few cases, the infants even left their mother to greet the stranger males and to watch what was happening to their mothers without being bothered by the attackers. The impression I gained was that the males wanted to impress female prisoners with their power and strength rather than harm them. An intriguing question is then whether such aggressive behaviour of the males affects the dispersal tendencies of young females before leaving their natal groups.

Following a classic approach to violence, I started by exposing here the males' side when encountering neighbours, but females as well showed an interest in neighbours. This females' perspective does not have to concur with the males' interest. As we will see, two aspects can be distinguished in the females. On one side, they fully support their males by being active participants in the males' violence. The observations of Zora, Sumatra and Wapi actively helping the males of the South Group to kill Aramis, as well as Zora eating his testes, perfectly illustrated this aspect. In addition, females can actively attack neighbours on their own. On 3 August 2005, at 8:56 hours, a large party of 14 adults of the South Group responded to drumming in the west with short screams and immediately moved silently in that direction. Four minutes later, while the adult males of the group had gone ahead to pursue the West Group neighbours, five adult females surprised a young female in a tree. Some of the females immediately climbed the tree and pulled her down to the ground, where they all attacked and struck her. She was prevented from moving by Sumatra. Three adolescent males, Woodstock, Utan and Taboo, tried to protect her, but the adult females continued with their attacks. After 2 minutes, Woodstock jumped on to Sumatra and hit her while she continued to impede the stranger. In response to this attack, Sumatra had to release her grip and the stranger immediately took advantage of the moment and escaped, pursued by the other females. In another instance, seven resident females of the South Group were aggressive towards a stranger female with an infant. However, three males of the South Group intervened to stop their attack such that, although the stranger female was injured and bleeding from cuts on the face and body, some of the attacker females were themselves hurt during this confrontation. Support from males of the stranger female's community then brought this attack to an end after 8 minutes.

In some instances, females seemed prepared to take great risks in attacking neighbours. On 17 August 2002, at 9:21 hours, Jessica, an isolated female of the Middle Group with her two offspring, was eating fruits in a *Scotellia coriacea* tree, when, to the west, she heard whimpers from a baby chimpanzee. Jessica made a face with a silent open grin, took her 9-month-old daughter on to her back and climbed down to the ground followed by her 4-year-old juvenile daughter. The baby moved to Jessica's belly and the juvenile daughter climbed on to her mother's back. After listening for 6 minutes, Jessica moved very carefully towards the west. After 17 minutes, she arrived under a *Sacoglottis gabonensis* tree where chimpanzees had been eating very recently. She smelled the fresh wads of food that were scattered on the ground and after 4 minutes started to eat some fruit for 25 minutes. Then, she climbed a tree with the infants, faced west and stayed, apparently listening, for 2 hours and 13 minutes. She then moved slowly and silently on the ground towards the west. At 13:18 hours, there was a short noise from a chimpanzee close by and Jessica's baby again clung to her belly while the juvenile got on to her back. She moved very slowly towards the sound and after 6 minutes, standing upright, she looked up into a *Uapaca* tree where two stranger females, one with a baby and the other with a juvenile female, were eating. Next, Jessica barked loudly and displayed towards the tree. The stranger females in the tree immediately screamed and rushed down disappearing towards the west. Jessica barked and screamed for 2 minutes and drummed once. The stranger females remained silent. Jessica barked five times during the following 20 minutes and advanced some 60 metres. At 13:53 hours, she retreated.

Jessica's behaviour was remarkable for two reasons. First, she was part of the Middle Group that at the time was already very small with only one adult male and two adult females, and therefore she could not expect much support. Second, being alone, it seemed as if she was counting on the surprise effect of her appearance in such a low visibility environment and, further, did not fear a counter-attack by the strangers after they had fled.

Besides being supportive to their males, female chimpanzees also visit neighbouring groups voluntarily, possibly in the search for appropriate sires. This is not in the interest of the males, but thanks to the fission–fusion structure of the chimpanzee society, it is impossible for males to prevent the females from moving on their own. Nadesh's visits to the South Group described in Chapter 2 illustrate situations in which visits to neighbouring communities are favoured by females. We have seen many such visits but since the females were not always habituated to human observers, our presence interfered with the visits so that they could not develop normally. However, the uniqueness of the situation in the Taï forest allowed us, through observing Nadesh, to highlight how systematic, intentional and important such visits can be for female chimpanzees. Although voluntary visits by young infant-less females before transfer into new groups have been reported in all the chimpanzee populations studied, voluntary visits to neighbouring males by fully grown adult females, with or without infants, have been documented more rarely. An extreme example of secondary transfers is provided by the mothers after the almost complete annihilation of males in the K group in the Mahale Mountains, Tanzania.[10]

A word of caution is needed here: females can be very shy and because of that we have much less knowledge of them compared with males. In Taï, it took 15 years before important and social females, like Salomé and Ella, tolerated our presence and allowed us to follow them during their daily activities. Discovering the social life of female chimpanzees can thus be especially challenging. In the Gombe chimpanzees, even after 32 years, there were still unknown mothers who appeared for short periods of time, whose social status was never clari-fied.[11] It was only after 20 years in the Taï project that we felt confident enough to say that all the females belong to one community and that they are all seen regularly. None of them has a peripheral status where they might be in contact with another chimpanzee community, as has been proposed for Gombe.

The summary of these observations illustrates the different interests of both sexes in neighbouring communities. Males attack and kill male strangers alongside mating with female strangers, while females attack female strangers and sometimes seek stranger fathers. Warfare in chimpanzee ranges from fatal violence against males and infants, to complex patrolling, or cooperative back-and-forth attacks and chases, to female prisoners kept for different lengths of time as well as peaceful visits to strangers made by females. Both males and females can play an aggressive role in intergroup attacks, while only females were seen to be involved in the non-aggressive aspects of such interactions. The competitive situation of the males does not, however, prevent them from having non-violent interactions with neighbour females; for example, Evered, one of the males in Gombe, was famous for having been seen in consortship with females from different communities. This complexity and mixing of aggressive and peaceful interactions is typical of the intercommunity relationships in chimpanzees. Some periods are more violent than others, and some periods seem more oriented towards sex. The demographic conditions prevailing between different communities may be the key element in determining the nature of interactions between chimpanzee communities.

IS WARFARE IN CHIMPANZEES ONLY ABOUT SEX?

My observations on the chimpanzees suggest that coalitions of males do not make war without considering the situation and evaluating the alternatives. Could we see signs that an evaluation of the situation is really taking place and that different factors affect this? We should not forget that warfare in humans is also a complex aspect of life as it is pursued to fulfil such a wide range of goals that experts still disagree about its exact function. If there is some agreement about the main causes of war in simple societies, it is that they are fought over subsistence, habitation or women.[12] Why do chimpanzees go to war? First, obviously the killing and emasculation of competing males clearly points to the removal of sexual competitors. The visits of fully adult females to neighbouring groups to have sexual interactions with stranger males also support a sexual interpretation of between-group interactions. At the same time, encounters happened repeatedly around fruiting trees and thus food competition might also play a role. My aim is to possibly identify the most important factors explaining the chimpanzee violence against strangers.

First, given that chimpanzees proceed to an evaluation of the situation before contacting the neighbours, their tendency to approach and attack them should be influenced by either demographic and/or nutritional factors. However, I was wondering if they really knew so much about their neighbours and if that knowledge influenced their decisions. To investigate the decision-making process about whether or not to attack, I analyzed data covering more than 13 years of encounters between the four known neighbouring communities that we followed simultaneously in the Taï forest. The Taï project is still the first one where such an analysis can be undertaken. I tried to see if a decision to go to war was based only on demographic factors within the aggressor group or whether it was influenced by their apparent perceptions of their potential opponents. The first surprise was that chimpanzees clearly knew quite a lot about their neighbours, as I could predict precisely if chimpanzees would approach or attack their neighbours only when including information from both the attackers as well as from the opponents. If including only information from the attacker side, I could not explain the decisions we observed.[13] Clearly chimpanzees have a detailed knowledge of their neighbours – but how precise is it?

Taï chimpanzees approach and attack stranger communities more often when they have relatively more males and fewer females than their opponents and live in lower density communities. The demographic conditions in the two opponent groups are important in deciding whether or not there will be a conflict. This knowledge of neighbours considers precisely the numbers of adult males, of adult females and the density in which the neighbours live. Furthermore, since we know that the Middle Group had four groups of neighbours and the South Group five neighbours, this means that chimpanzees roughly keep track at any one time of the composition of the communities of several neighbours. As we know neighbours to be sometimes of dramatically different size and composition, the chimpanzees seem to associate each neighbouring group with different territory locations and specific demographic characteristics. Finally, they compare the demography of neighbours to their own one, so that one day they may have more males than one of the neighbours, while the next day it may be exactly the opposite with another neighbour. At the same time, demography is not fixed in stone, so that over time this may change for a same opponent group. Given that contact between males of neighbouring groups is infrequent, very aggressive and brief, we are left wondering how exactly they gain such precise knowledge.

Second, if the number of males is important, at the same time chimpanzees are willing to attack groups with more females than they have themselves, even if they have also more males. As an illustration of the decision process and why chimpanzees might decide to attack more powerful opponents, let us take the example of the Middle Group in the Taï chimpanzee project. The Middle Group was always the smallest of our study communities with only twelve chimpanzees when we first habituated them. It was impossible for us to know why this was such a small community, but it supported our impression that community size in wild chimpanzees varies greatly and small ones do exist.

The Middle Group always had a relatively high number of adult males: in 1996, there were four adult males for three adult females with four youngsters. At the same time, the North and South Groups had three times more females, but only the South Group had more males (seven adults and adolescents). Despite being the smallest community in the forest, the Middle Group was the most active of the three groups in terms of incursions into neighbouring territories and seeking physical contacts: they made 40 such incursions in 2000 and 21 in 2001. This was two to three times more than the South or North Groups.

Males of the Middle Group took many risks to find females. If we expected them to seek new sexual encounters with the least risk, we would have expected them to attack the North Group. However, the Middle Group sought out the South Group almost twice as often as the North Group; in 2000, there were eight encounters with the South Group and four with the North Group, and in 2001, nine with the South Group and six with the North Group. During these 2 years, the South Group had, respectively, five and six males for 20 and 18 females, while the North Group had two males and seven females for both years. So, the Middle Group chose the most powerful neighbours but, at the same time, chose the group with three times more females than the others, and six times more than they had at home. During these 2 years, we saw the Middle Group males capture temporarily one female from the North Group, two females from the South Group, as well as four other unidentified females, two of them with babies.

In 2002, the Middle Group lost its old experienced male and the number of encounters with neighbouring communities decreased dramatically to 12. Nevertheless, they were still seen twice with a stranger female. Sadly, in 2003, another male in the Middle Group died of disease and the remaining one then avoided encounters with his neighbours.

The Middle Group history illustrates well how chimpanzees are not afraid to attack larger and stronger communities, as long as the number of females in the opponent group makes the adventure potentially attractive. With the precise knowledge they seem to have of their neighbours, they can evaluate the potential benefit and therefore seem ready to attack stronger opponents when enough females are present. The frequent mating with female prisoners gives a direct benefit to such attacks from the male's point of view. It is more difficult to evaluate the long-term impact of such behaviour. Do aggressive males gain a reputation among the females that will influence their future transfer choice? Will young females be more inclined to migrate later into their attacker's community or, on the contrary, avoid it?

Killers might be rewarded by attracting females. The example of the South Group suggests that this may be the case. The group had become exceptionally aggressive since the young males, Sagu and Gogol, had gained the two highest positions in the hierarchy. It was the conduct of these two ambitious males that started the South Group's terrible violent encounters with their neighbours. It was the first time in 25 years of the Taï project that we had had this pairing of young dominant males, whereas up until then there had only been prime males, around 25 years of age, taking up the alpha position. We are certain that it was Gogol who killed the infant from an unknown stranger community, a few months after he became the beta male. Similarly, it was Sagu who killed Aramis of the East Group. In addition, both were present when Nérone of the East Group was killed, when the two other adult males present were elderly and whose teeth were too badly worn to be able to inflict such wounds. It is especially relevant for this discussion that at the end of January 2007, for the first time since 1995 when we had started to follow the South Group, two new immigrant females integrated into the community. Thanks to genetic analysis we could be precise about the history of the second female: she immigrated first in June 2006 in the Middle Group, before transferring a second time, in January 2007, into the South Group. Seven months later, Jessica, one of the remaining mothers of the Middle Group, transferred permanently into the South Group, where she was accepted without much violence and her son Jonathan, three and a half years old, was unharmed and rapidly seen playing with the other youngsters. Furthermore, as seen in Chapter 2, Nadesh, the second remaining female of the Middle Group, when in oestrus, repeatedly visited the South Group. Were these females attracted by the strong young males of the South Group? Difficult to draw firm conclusions, but the coincidence is informative.

How can chimpanzees gain such a precise knowledge of the number of chimpanzees in neighbouring communities? My feeling is that numerical assessment of neighbours is obtained through a detailed vocal recognition of individuals. Vocal exchange with a lot of drumming is always part of the interactions between communities. Voice recognition is clear within communities; could it be that it also exists between communities? Playback experiments performed with chimpanzees in the Taï forest showed that they not only recognized the voices of their male neighbours but also that they react differently if they hear them within their territory limits or outside of them.[14] For example, Middle Group males showed no special reaction to playback calls of their own group members, whereas they responded aggressively to playback calls from North or South Group males. Furthermore, they reacted clearly more aggressively when there were calls from more than one individual. Intriguingly, they reacted differently when, for example, the South Group male voices were played from within their proper territory than from within the North Group territory, suggesting that they also knew the location of their different neighbours' territories. We confirmed that chimpanzees could differentiate the voices of males of their own community from those of neighbouring ones.[15] Whatever the precise mechanism, chimpanzees seem to remember exactly the number of males in neighbouring groups and compare this assessment with the number of individuals in their own group to make decisions about when to attack.

For females, such assessment seems more difficult as they do not call systematically during encounters and also are present less regularly. However, it is obvious that more visits must occur between communities than we actually observed, since we could not always be present. This seems to be enough for them to gain a reliable indication of the number of females in the different neighbouring communities. To gain an idea of the densities of neighbouring groups, it seems that they are able to use their knowledge about the location of the neighbours' territory. Knowledge of range and size has to be built up over time as encounters are repeated with the same communities. This will lead to a sufficiently accurate estimate of density, which is indicative of feeding competition. All this suggests a precise and long-term memory-based knowledge of neighbours in chimpanzees.

We conclude from this analysis that sex is not the whole story: elimination of rivals, sex and food competition as related to density all coincide to make chimpanzees attack their neighbours. Chimpanzees take into account the two main driving forces for survival and reproduction to make decisions about conflicts. This complex decision-making process is not only made by old experienced warriors, like Brutus, but also by other males when they hurry to support their group members whenever the circumstances require it.

Raiding the weakest possible enemy is a preferred human strategy. Some have proposed that chimpanzees do the same. However, in Taï, the very cooperative dimension of intercommunity chimpanzee violence makes the number of males within the whole community more important than the sheer number of males at the moment of an encounter. The systematic dimension of support seen in Taï changes the rules of war, since the forces present are a poor indicator of the forces actually available.

Generally, it remains that chimpanzees follow different purposes at the same time. So that if it is the elimination of rivals that counts, then the number of males will be critical. If, however, sex is the driving force, then the situation becomes more complex as the relative numbers of both females and males have to be considered. The numerical assessment done by chimpanzees before contacting their opponents follows a precise evaluation of the forces in presence. As this is done before visual contact, it has to be a mental evaluation based on precise memory of neighbours.

PARALLELS BETWEEN CHIMPANZEE AND HUMAN WARFARE

Chimpanzee wars include a complex set of behaviours ranging from the taking of prisoners to killing rivals and sex. These aspects in humans and chimpanzees are compared in more detail in Chapter 7. For now it is important to note that war, when defined as 'socially organized conflict between members of different territorial units' is observed in chimpanzees. Scenarios which consider the evolution of war need, therefore, to take into account the possibility that many aspects of this behaviour were present in our common ancestors and that neither agriculture nor fixed settlements, nor the use of tools or weapons, were necessary for the emergence of war, although they certainly affected how it has developed and the decision-making process.

Why did war develop only in chimpanzees and humans? In many other species, we see aggressive males defending territories and going as far as killing intruders. In a lower number of these species, we see groups of males fighting to defend their territories and killing neighbouring males or invaders killing group leaders. Lions, hyenas and wolves are clear examples of this, all of which show strict, communal territorial defence behaviour. This has some similarities with humans and chimpanzees, where, however, territorial defence is undertaken by comparatively larger social groups. Unlike these carnivores, both human and chimpanzee males usually remain in their natal group throughout their lives, and are subject to important predation pressure by large carnivores, thereby developing a strong sense of group belonging. Under these circumstances, xenophobia can develop, whereby outsiders are violently ostracized and killing them seems to have a different value than does killing a fellow group member. A second factor is important here and that is sexual competition. Due to the increase in maturation time in the hominid line, females spend most of their adult lifetime in maternal investment and so, for the same number of females, male chimpanzees and humans have much less chance of reproducing than do carnivores, for example. This leads to strong nepotism in large close social groups and to competition between neighbouring groups for mating opportunities. In other words, I suggest that the preconditions for warfare are high predation pressure leading to larger social groups in species with a slow reproductive rate.

The ambivalence mentioned above between competition and cooperation in male chimpanzees lies partly in this within/outside distinction. Much of the competition is exacerbated between communities, whereas cooperation dominates within communities. So, contrary to hypotheses that war is the result of a genetic drive in both chimpanzees and humans, leading to violence between males,[16] I think rather that war results from strong cooperation between males within communities. This leads to strong nepotism and a lower regard for outsiders. Raymond Kelly speaks of a social substitution process in humans, whereby the killing of outsiders is glorified while that of group members is punished as not morally justifiable. Such a process in human societies only becomes possible when there are sufficient levels of complexity in kinship rules. In chimpanzees, it is the systematic occurrence of cooperation and reciprocal altruism within communities that leads to increased solidarity and strong feelings of belonging and, its antithesis, xenophobia. This results from the daily acts of solidarity between members of a community and not from a formal categorization.[17] As an ironic twist, this increase in solidarity led to a stronger aversion towards male outsiders and made extreme violence possible. However, this remains sex specific, as it is addressed mainly against males, the sex rivals, and not against females, the object of this competition.

Therefore, the observations made on chimpanzees suggest that war could have appeared much earlier than when specialized weapons appeared in human evolution. It has been proposed that human ancestors lived for a long time in conditions of high predation pressures, the classic image being that of man being hunted by sabre-toothed tigers and lions. In addition, it has been suggested that our *Australopithecus* ancestors had long generation times with slow maturation and a heavy investment in offspring. Both are conditions we see in chimpanzees, and I consider them as conducive to the appearance of warfare.

In chimpanzees, the intensity of war can be very diverse. The North Group has been followed for 28 years and has never shown any tendency to kill strangers. They have demonstrated all the panoply of warfare tactics as seen in other groups, but even with a large group of 10 adult males and six adolescent males to support them, they were never seen to commit such violence. We tended to think that some aspects of life in the Taï forest made the killing of strangers unnecessary. The first 10 years of observations with the Middle and South Groups confirmed this impression. It is only since April 2005 that we have known that the killing of conspecifics is part of warfare in Taï chimpanzees. In other words, warfare with its complex fighting strategies can be conducted without extreme or fatal violence. From our long-term observations, chimpanzees seem to pursue some specific aims when fighting strangers and these can be achieved often without killing.

It is important to bear in mind that in humans also the intensity of warfare can differ dramatically between societies. Some societies have been described as peaceful as they rarely go to war, while others have been described as fierce, because their involvement in warfare is more or less continuous. In general, the intensity of war has been proposed to increase under three different conditions.[18] First, warfare is expected to develop in rich environments in which comparatively high population densities are found over extended periods of time. Secondly, it was proposed that in many human societies, war is about marriages and exchange of women, and the more exchanges, the more conflicts can occur, leading to more wars. Thirdly, many prehistoric intensive cases of warfare seem to be associated in regions with difficult times due to ecological and climatic changes.

This illustrates that under different conditions, the state of war is also in human societies influenced by specific conditions that make war profitable or obligatory to survive or preserve favourable reproductive conditions. In a way similar to some human groups, the chimpanzees in the Taï forest experience situations where attacks against their neighbours can be the best way to improve their prospects for sexual interactions and normal reproduction, while at other times they seem much more peaceful and avoid fatal confrontations for prolonged periods of time. Communality of demographic conditions and in ecological constraints could have led to the development of warfare in both chimpanzees and humans, both developing it further in different ways. We will come back to this in Chapter 7. Before then, I would like to return to the regular mention I made during this and the previous chapters concerning the differences between chimpanzee populations and concentrate on these in Chapter 6. There, we will see in detail and try to understand how different ecological conditions and different levels of large carnivore predation faced by chimpanzees affect their behavioural diversity. This will stress one of the most important findings on chimpanzees of the last three decades, namely the extensive flexibility prevailing in this species.

NOTES

1. These are the words used by the specialists Ember and Ember (1992). War specialist Lawrence Keeley talks in his book *War before Civilization* (1996) about 'armed conflicts between different groups of humans'. Keith Otterbein in his book *How War Began* (2004) defines war as 'armed combat between political communities or independent groups' with a focus on weapons and not aggression. He further distinguished seven forms of killing: homicide, political assassination, feuding, warfare, duelling, capital punishment and human sacrifice. Raymond Kelly proposes in his book *Warless Societies and the Origin Of War* (2000) a more detailed definition of war that entails 'armed conflict that is collectively carried out', and is an 'organized activity that required advanced planning' and where the 'use of deadly force is seen as entirely legitimate by the collectivity'. He proposes on page 140 to distinguish between three types of aggressive attacks in human warfare, ranging from spontaneous armed conflicts over resources, to revenge-based killing culminating in the punishment of indiscriminate members of the offenders. For him, this kind of social substitution of the revenge principle is the hallmark for war. However, his limitation of war to revenge cases might be too restrictive and does not agree with other specialists.
2. Within-group killing of adults has been seen in only two instances. One young male was killed by his male group members in the Ngogo chimpanzee community, which is abnormally large with over 25 adult males and where competition between the males seems very high (Watts 2004). Another case has been witnessed of a young male killed in Budongo chimpanzees (Fawcett and Muhumuza 2000, Reynolds 2005).
3. In our book *The Chimpanzees of the Taï Forest: Behavioural Ecology and Evolution* (Boesch and Boesch-Achermann 2000), we presented our observations suggestive of a possible difference in the use of violence in intergroup relationships. This illustrates the difficulties of drawing conclusions from observations on behaviour patterns that are rarely seen or difficult to interpret.
4. Male lions emigrate out of their natal group to live for some years in bachelor groups before attempting to take over prides of females. Thanks to genetic studies it was possible to show that large coalitions of males, which are more successful in controlling a pride for long time and producing a large number of cubs, were all composed entirely of close kin (Packer *et al.* 1991). Coalitions of two to three included only non-related males and shared reproduction much more evenly.
5. The development of war activities has been documented in many ethnographic descriptions in vivid detail (Godelier 1984, Chagnon 1988, Descola 1993, Keeley 1996, Gat 1999, 2000, Kelly 2000, Otterbein 2004, Arkush and Stanish 2005). They showed that war strategies are very diverse, in particular about how and when to attack, but there is a clear and strong difference between developed state societies, where armies are present and large groups of warriors face one another, compared with non-state societies, where raids carried out by small groups of equal men are the main activities.

6. All the observations have been made following pre-established protocols by a very efficient and motivated team of local field assistants who have worked on the projects for many years, as well as different students that contribute to the general data collection. The analysis of intergroup encounters, therefore, represents a large collaborative effort (Boesch *et al.* 2008, in press a). For more detailed descriptions and most quantifications, I use my and Hedwige's observations where we followed 485 encounters between two neighbouring groups during a 13-year period (Boesch and Boesch-Achermann 2000). Of these encounters, 25% resulted in a visual contact between the two opponents, when at least one of the two groups approached the other. Furthermore, 10% of all encounters resulted in physical contact between the two opponent groups. They patrolled their border in 38 out of 129 times on their own initiative, while they reacted to the sound of strangers on 91 occasions.

7. Decreasing group size is personally a dramatic process to follow, as most individuals about which we are getting more information, and sometimes that we have been following for over 20 years, are disappearing. On a purely scientific point of view, however, such a decrease mimics an experiment where you would follow the reaction of the group after removing a different number of individuals. This allowed us to determine precisely within the same group the role of demographic factors on the conditional use of differential territorial as well as hunting strategies (Boesch and Boesch-Achermann 2000). In both contexts, we could follow how chimpanzees adapt their behaviour to the current demographic situation prevailing within their social groups.

8. Infanticide in stranger females has been observed in different populations and remains the most common form of intraspecific killings between groups of chimpanzees in Gombe, Mahale and Ngogo (Goodall 1986, Nishida *et al.* 1990, Watts and Mitani 2000). In strong contrast to these populations, infanticide was seen only once in Taï chimpanzees and that despite numerous observations of stranger mothers being with neighbouring males (41 cases, Boesch *et al.* 2008).

9. In no other domain than the one of observing non-habituated chimpanzees is our presence as human observers less welcome. Wild chimpanzees react with sheer panic at the sudden appearance of humans in most situations, and so our presence can terribly affect the outcome of any encounters between the habituated study groups and a non-habituated one. Therefore, we tried to be as discreet as possible and remain at a distance where we might not be seen or be less threatening. This, however, makes detailed observations more challenging and incomplete.

10. The K group in Mahale Mountains National Park, Tanzania, decreased dramatically in the number of adult males and then, over a period of three years, at least six females, some with infants, started to migrate to the larger neighbouring groups to the point that after a few years only one male and two females remained from the K group (Nishida *et al.* 1985).

11. The status of females has been given more interest recently and a much more complicated situation than previously presumed has been revealed (for Gombe: Williams *et al.* 2002; for Ngogo: K. Landgergraber *et al.* unpublished data; for Goualougo: C. Sanz *et al.* unpublished data). In Gombe chimpanzees, out of 22 females, three were given a peripheral status as it is possible that they lived their adult lives in more than one group.

12. The function of war seems to be clouded by many disagreements. Otterbein (2004) sees war originating with the hunting of large mammals when the first

possible hunting weapons appeared with *Homo habilis* and then war disappeared again as mammals became extinct due to overhunting by early hominids. Much later war reappeared with the development of agriculture, and fixed settlements. Thus, war begins with *Homo habilis*, and not in chimpanzees, as they do not use weapons. For Kelly (2000), warfare develops much later in rich environments in which comparatively high population densities were sustained over extended periods of time without possible means of escape. For him, it requires segmented human societies, with extended kinship rules that are absent in hunter-gatherer groups. Population density has been challenged as a reason for producing war, as the likelihood of going to war every year is the same for populations with fewer than one person per square mile as for those with 100 persons per square mile (Keeley 1996). Manson and Wrangham (1991) proposed that 'inter-group conflicts will be over resources when alienable resources are available, as these are the base for survival, otherwise it will be over women'.

Revenge for earlier killings, stealing a woman or trespassing is often presented in interviews as the main causes of warfare in many societies (Keeley 1996, Kelly 2000). For example, in Native North Americans, war originated in 93% of cases as revenge for killing, 60% in retaliation for poaching and 58% for the capture of women (the total percentage exceeding 100% indicates that people give more than one reason for the cause of war). Since revenge is not an explanation in itself, it simply illustrates how obsessive humans can be in situations they consider important. So that in this example, poaching and the capture of women leads to war and if not settled satisfactorily revenge will continue to be sought.

13. When analyzing only the intergroup interactions between neighbouring groups in the Taï forest where we knew perfectly the composition and the territory size of both opponent groups, we were able to study the most important factors affecting the decisions of the Taï chimpanzees (Boesch *et al.* in press a). When including all the different possible factors that could affect the decision of the attackers, such as the number of adult males and females present in both communities, the number of sexually active females, the sex ratio of adult members as well as the size of the different territories at the time of the encounters, we could see that when taking in account only factors from the attacker's point of view we were unable to explain the likelihood that attackers would approach and come into contact with opponents. If, however, we considered the perspective of both the attacker and the opponent groups, we could explain precisely when attackers would approach and contact opponents.

14. In a detailed study on the knowledge of neighbours, Ilka Herbinger used playback experiments with the Taï chimpanzees. By recording a single voice or choruses of known males from three neighbouring communities and from one stranger one, she could, with the help of powerful loudspeakers, play the recordings at predetermined locations to simulate the presence of strangers. A similar numerical assessment as in Taï has also been shown with chimpanzees in the Kibale forest of Uganda (Wilson *et al.* 2002).

15. With precise spectrographic studies, Catherine Crockford showed that the voices of males within a community were more similar to each other than to the voices of males of the directly neighbouring communities (Crockford and Boesch 2004).

16. Richard Wrangham, in his works on the 'Demonic Male', stressed a natural propensity for lethal forms of aggression in male chimpanzees and humans that lead to intercommunity killings (Wrangham and Peterson 1996,

Wrangham 1999, Wrangham *et al.* 2006). He postulated there a section of the male psyche that seeks opportunities to carry out low-cost attacks on unsuspecting neighbours that would be accompanied by an enjoyment of the chase, a victory thrill, a tendency to treat non-group members as prey. A dominance drive by male groups that is favoured by selection is postulated here.

17. Keith Otterbein (2004) viewed 'fraternal interest groups' as the requirement for the development of war. This idea made him also view war as something that developed very early in the evolution of humans. However, he related it to the development of hunting of large mammals with weapons. Using his terminology, I suggest that the fraternal interest groups are both necessary and sufficient for the development of war. Weapons, once acquired, would certainly be rapidly used and lead to a new development in the conduct of war.

18. Kelly's (2005) analysis supported the conclusion that both environment richness and population densities affect the likelihood of war. Ericksen and Horton's (1992) analysis suggested that exchange of women was an important driver of war frequency. Finally, Ember and Ember (1992) proposed that important ecological changes were often seen to force human groups to start war.

6

The real chimpanzee

The 'F' family is the most famous dynasty of chimpanzees from Gombe, as Flo, the old and successful matriarch, and her 5-year-old daughter Fifi were made world famous thanks to Jane Goodall's book *In the Shadow of Man*. The daughter, Fifi, became the most successful mother among the Gombe chimpanzees and her two oldest sons were long-lasting dominant males in the group. Frodo, the second, was the most eager and brave hunter when I followed them in the early 1990s to understand the difference in hunting strategies from those used by Brutus and consorts in the Taï forest. Frodo was also physically outstanding as one of the largest and heaviest males that ever lived in Gombe. Because of his status as the keenest of the hunters, I followed him most of the days during my stay in Gombe. One day in April, while with four chimpanzees, we heard Frodo screaming in the direction of a group of red colobus monkeys. We were high up on the hills, where we had a perfect view of the whole valley and the Tanganyika Lake glowing below us in the sun. The dominant male, Wilkie, charged under the colobus monkeys followed by the others. The group of about 20 monkeys was isolated in a large tree in the grassland with, as their only escape route to head back into the forest, two smaller trees side by side. My Taï chimpanzee eyes recognized a perfectly easy hunting situation. However, Frodo continued screaming, as he climbed two metres in the tree of the monkeys and looked at them without approaching, whether out of fright, I don't know. The other chimpanzees just sat, waiting and looking at Frodo. Immediately, two large male colobus, balls of fur with all hairs on end and aggressively screaming, charged at Frodo. After 10 minutes, during which Frodo continuously screamed, one of the male colobus jumped on Frodo who then rushed to the ground, attacked simultaneously now by four of the males. To my Taï chimpanzee eyes, an unbelievable chase ensued, where the four male monkeys pursued screaming Frodo through the grassland towards the forest, while the other chimpanzees looked on without doing anything, some seeking refuge in the small trees. Then, all came back to the same tree, with Frodo still wanting to hunt without daring to do so. He was attacked two more times by a male colobus, twice having to shake him off his back. In the end, the colobus group, including all the females, started to move back into the forest. What a mistake – that was exactly what Frodo had waited for! He manoeuvred himself through the monkey males higher up to grab a baby from a mother's belly. As he did so, two other chimpanzees used the arising panic amongst the monkeys and grabbed two more babies away from their mothers.

So, in all a very successful hunt, but how different from what I was used to observing in the Taï forest! The savannah, making the cornering of monkeys so much easier, the fear of the male colobus that was blocking Frodo, something I never saw in Taï, and the absence of cooperation and support among the chimpanzees that merely looked at Frodo fleeing from the attacking monkeys were all new to me. How could the interactions between chimpanzees and colobus monkeys be so different at the two sites? Obviously, the small tree size and the patchy savannah habitat made the hunting conditions very different and favourable for catching prey in Gombe compared with the Taï forest.

However, not only the hunting is influenced by the ecological conditions prevailing under which chimpanzees live and which can lead to a cascade of effects. I suggested in the first chapter that the importance of ecology has not always been fully recognized, for the reason that when our research on the chimpanzees began, we had extremely sparse knowledge about them. Then, as understanding developed, studies were concentrated mainly on one population: the notion of population differences is a relatively new one. To emphasize this development, I will start by looking at the discovery of the chimpanzee and track time backwards to give an overview of how our knowledge about this species evolved.

FROM SATYR TO OUR CLOSEST COUSIN

It took a long time for chimpanzees to be recognized as a species and surprisingly longer for them to be accepted as humans' closest living relative. Although the process started as early as 470 BC when they were first documented, this first observation was attributed to another species. At that time, the Suffete Hannon organized a commercial expedition with 60 galleys of 50 oars from Morocco to Upper Guinea to establish new colonies. In the foothills of the mountains of Serra Léôa (Sierra Leone), the people from Carthage met some *Gorillai*, hairy inhabitants of the forests that defended themselves by throwing stones at the sailors. Pliny reports that two skins acquired on that occasion were still preserved in the temple of Astarté in Carthage at the time of the Roman invasions in 146 BC. At the same time, an old mosaic on the floor of the temple of the Fortune in Préneste (Palestrina) illustrated some 'satyrs' of the Upper Nile. There then followed a long period of silence. We had to wait until 1598 for the Portuguese to mention, through the voice of Philippe Pigafetta, the presence of these animals in regions of Congo or Angola. They were called in turn, *Satyrus indius*, *Homo sylvestris* or even mandrill! The British scientist E. Tyson was the first to study their anatomy, in 1699. In 1740, the count George Buffon obtained the first live individual of the species. In 1762, in his *Natural History of the Quadrupeds*, Buffon describes 'a monkey as tall, as strong as Man, as passionate for females as Women, a monkey that can carry weapons, that can use stones to attack, and clubs to defend himself and that resembles Man more than the *Pithecus* [gorilla and orang-utan in the terminology of that time], as he has a kind of a face with traits close to Man'.

In 1758, Carl von Linné, the father of modern taxonomy, published his *Systema Naturae*, and proposed a modern system of classification for all animals based on species, families and orders. The Pongidae was separated from the Hominidae and included creatures as curious as *Troglodytes bontii* or *Homo troglodytes*, active at night and expressing themselves by whistles, *Lucifer aldrovandii* or *Homo caudatus*, possessing a tail, *Satyrus tulpii* (the chimpanzee) using tools or *Pygmaeus edwardi* (the orang-utan).

At the beginning of the last century, the debate around the Piltdown man in 1912 emphasized the fact that our ancestors had quickly acquired a modern human type of brain and, therefore, the great apes were less human than previously assumed. The relation proposed between humans and the African great apes by the British naturalists Charles Darwin and Thomas Huxley and the German Ernst Haeckel fell out of fashion. The fact that the Piltdown man was revealed to be a forgery in 1953 did little to change the attitudes of anthropologists towards the great apes. It was not until the 1980s that reason prevailed and genetic data forced the scientific community to see the chimpanzee as man's closest living relative.

FROM MYTH TO REALITY: THE DEVELOPMENT
OF CHIMPANZEE STUDIES

Since the early 1900s, chimpanzees have been studied mainly in captivity. One of the pioneers, the psychologist Wolfgang Köhler, set up a
laboratory for captive chimpanzees near Conakry in Guinea, and performed some innovative studies with chimpanzees. He was the first, in
the 1920s, to show that in problem-solving situations his chimpanzees
were capable of 'insight', that is, able to solve problems by mentally
considering alternative solutions and then performing the correct one.
It took over 60 years for his colleagues to accept his conclusions. Later,
at the Yerkes National Primate Research Center, Atlanta, USA, studies
were also carried out with captive chimpanzees. Sadly, chimpanzees
were used mainly for biomedical research and in these facilities the
living conditions were determined more by the medical needs and not
by the well-being of the animals. It is only during the last two decades
that captive settings slowly started to take account of the biological
needs of species held in captivity and the living conditions for these
animals were improved.

The first short field studies on chimpanzees started in the 1930s
and remained anecdotal until 30 years later, when the first long-term
project on wild chimpanzees in their natural habitat started: in 1960,
Jane Goodall was the first to initiate a long-term project on the chimpanzees of the Gombe National Park in Tanzania on the shore of Lake
Tanganyika. She was quickly followed, in 1963, by a Japanese team led
by Toshisada Nishida that started a project on chimpanzees in the
Mahale Mountains National Park, also on the shore of Lake Tanganyika,
some 200 kilometres south of Gombe. Both projects are still active and,
for the first time, are able to adopt a generational perspective on this
long-lived species.

In 1979, the project in the Taï National Park, Côte d'Ivoire,
started. It was the first long-term project to study chimpanzees in a
dense tropical rainforest with the aim of learning how ecological conditions affected the behaviour of this species. At about the same time,
a Japanese research team led by Yukimaru Sugiyama started a project
at Bossou near Mount Nimba, Guinea. In the late 1980s and early 1990s,
three new long-term projects began in Uganda and, very recently, a
project to study the chimpanzees of Central Africa started in the
Goualougo Triangle in the Nouabele-Ndoki National Park in northern
Congo. The Central African chimpanzees remain the least studied by

far of the various populations, although they probably still outnumber chimpanzees from other regions. This bias might be due to the fact that all the Central African populations are forest dwellers and working in the forest has a reputation for being logistically difficult, or because they are in French-speaking countries, while most of the scientists working on primates originate from the English-speaking world. Whatever the reasons, this situation leaves important gaps in our knowledge about one of the largest populations of chimpanzees.

WHAT IS A TYPICAL CHIMPANZEE?

I really do not like this question as I think that there is no correct answer. However, people do not like this reply; scientists even less so. There are two obvious ways of answering this question: the first is to say – 'mine'! One's first experience with a population of wild chimpanzees is very impressive and tends to have a lasting effect on the scientist involved. Logically, one would then assume that all chimpanzees behave in the same way. This may be more subtly put in that one judges the appropriateness of results in the light of personal experience, so that observations that do not accord with personal experience are greeted with scepticism. The second way of answering the question is by saying that the most famous chimpanzee population is the standard. In which case, the Gombe chimpanzees are taken by most people as 'the chimpanzee', and differences from this standard have often been viewed as deviations rather than important additional information. If one looks at the history of why the chimpanzees of Gombe were first selected for research, this is understandable but has also turned out to be misleading.

Louis Leakey, the famous British palaeo-anthropologist who discovered *Australopithecus boisei* and *Homo habili* in the Olduvai Gorge, was convinced of the savannah model, by which our human ancestors gained their human features once they ventured into the open savannah of East Africa. To confirm that idea, he sent young Jane Goodall, his assistant, to study the only chimpanzee populations he knew to be living in the same conditions as those faced by our human ancestors. His idea was that there, chimpanzees, being our closest living relative, would have developed so-called human abilities, such as tool use and manufacture, hunting behaviours and others. If so, this would support some aspects of human evolution theory.

Leakey's intuition proved to be correct, as it took no more than 3 years for Jane Goodall to publish revolutionary papers in the internationally renowned scientific journal *Nature* revealing that wild chimpanzees spontaneously make and use tools, and that they hunt monkeys for meat. She reported chimpanzees making small fishing tools from bark or small branches to extract termite soldiers from large termite mounds during the swarming season. In some instances, these tools were made before reaching the mounds, indicating some anticipation of future activity.[1] The whole scientific community was shaken by such revelations and the chimpanzees gained the status of confirmed close cousins of modern humans. A discussion unfolded about the reality of tool use in chimpanzees and how it compared with humans and this debate is still ongoing, but it placed the chimpanzee in a position no one had predicted.

The incredible success of this project resulted in the Gombe chimpanzees being taken as typical of the whole species. Still today, after 40 years, the Gombe chimpanzee is the best studied population, the one with the longest list of publications, and by far the largest cohort of students that have become accomplished scientists, while Jane Goodall is the best known living scientist on the planet. So it is not too surprising that, for many, the chimpanzee of Gombe stands for 'the chimpanzee' and any difference in behaviour found in other populations is often considered to be a deviation from the standard. However, Louis Leakey intentionally selected a population of chimpanzees living in a marginal habitat, a habitat that best paralleled the conditions of our early ancestors, and not a habitat that is representative of that of most living chimpanzees in Africa. All our present knowledge on chimpanzee distribution shows that the majority are found in the rainforest from Sierra Leone to the Democratic Republic of Congo. It is true that they have been able to adapt to drier habitats in Senegal, Mali, Nigeria, Uganda and Tanzania. However, these populations are much smaller than those found in the forested regions, where by far the majority of chimpanzees have always lived.

A posteriori, Louis Leakey was wrong: the fact that the chimpanzees in Gombe hunt and use tools does not support the theory of human evolution as originally thought, but simply revealed to the scientific community the amazing achievement of this species. Tool use and hunting are common to all chimpanzees. The assumed effect of the environment, namely that adaptation to a more open habitat favoured the evolution of tool use, tool manufacture and hunting, could have been made only at a time when we had no knowledge about the behaviour of chimpanzees living in rainforests. Absence of evidence is not evidence of absence. This is one of the general problems of science in that we ignore what we do not know and, therefore, it is more sensible to refrain from generalizing and rely mainly on positive results.

Thus, from all I presently know, I propose that chimpanzees were originally inhabitants of the forest, where they acquired most of their abilities, and that only later they adapted to more open and dry habitats. However, most of their adaptations are forest adaptations and some of these abilities might then have been lost or altered when adapting to new habitats.

IT IS A TOUGH LIFE IN THE FOREST

The daily challenges of life in the forest have shaped chimpanzee intelligence and social behaviour. One important challenge is to find food. Food is plentiful but not always easy to access. For example, insects, such as ants, bees, termites, beetle grubs and caterpillars, are present everywhere but most are either hidden underground or in nests inside tree trunks, branches, saplings or other vegetation. Therefore, either brute force, such as used by gorillas and orang-utans, is necessary to obtain them, or tools have to be invented that will enable access to the hidden food. Selection for inventing tool use was stronger as more profit could be achieved from it for chimpanzees. The forest is a very rich habitat, especially in terms of the amount and diversity of edible insects, and the exploitation of this rich food niche is very beneficial. Alternatively, an open habitat, as in the Gombe or Mahale National Parks, may look closer to the living conditions of our ancestors, but from a chimpanzee's point of view seem to contain fewer and less diverse insects. Taï chimpanzees eat insects on a regular daily basis, whereas they are eaten less frequently at Gombe and Mahale.

In April 1985 in the Taï forest, I followed Héra and her two sons, Haschisch and Eros, while they foraged quietly in the undergrowth, looking for food. At 08:00 hours, they arrived at the large *Parinari exselsa* tree they had been visiting for the last few days. The ground was covered with a layer of nuts and Héra gathered some in both hands and settled at a large anvil with a granite stone. The youngster Eros, about 2 years old, was already there, facing her and ready to eat. He begged for nuts from the start and, as usual, his generous mother let him have many of the nuts she pounded. Regularly, after opening a nut, Héra took a small 12-centimetre long, hard stick, sharpened it with the teeth and used it to extract the almond remains in the nut shell that she was not able to reach with her teeth. Héra was a very skilful nut cracker and she opened them with just three to five strikes, thereby eating on average one and a half nuts per minute. Haschisch, 6 years old, had also settled under the same tree at another anvil with a stone that he had used the day before. A heavy stone is a great help in opening the hard-shelled *Parinari* nuts. It is possible to pound them with a wooden club but it takes much more effort. Haschisch had two fingers of each hand paralysed but was nevertheless a keen and determined nut cracker. He did not beg and pounded, all by himself, eight to ten nuts at a time before gathering more, and he had an efficiency rate of 12 strikes per nut, eating fewer than one per minute – thus still a beginner in comparison with his mother. Despite his young age, Haschisch worked that day just like Héra, without interruption, for 5 hours and 15 minutes.

Later, Héra found a nest of sweat bees in a freshly fallen large branch. The nest was hidden inside the wood, its small entrance betrayed by a few swarming bees, but most of the bees must have already abandoned the nest. Héra quickly broke off a stick from a sapling and reduced its length to 25 centimetres by severing it with her teeth. She dipped one end of the twig deep into the hole and rapidly withdrew it with pieces of drying honey sticking to the tip. She licked off the honey and dipped for more. Again, Eros sat nearby. When she withdrew the twig, he placed his fingers tightly around it and she pulled it through his fist so that he could lick the honey that remained on his palm. She continued to dip and soon started to collect larger amounts of honey. Eros now begged directly for the loaded tool. Héra handed it over to him and quickly made a new one. They now exchanged tools – the empty one for the loaded; she dipped while he licked. Héra shared every second dip and this went on for about 4 minutes.

Taï chimpanzees possess a wide range of tools – cracking nuts with stone hammers, extracting honey from beehives with long sticks, fishing for ants with medium-sized sticks, extracting grubs and honey from wood-boring bee nests with small twigs, using leaf sponge to drink water out of tree holes, extract bone marrow with small sticks, etc. – and are regularly seen to make tools before using them. When comparing their tool repertoire with those known from Gombe and Mahale chimpanzees, the conclusion is that Taï chimpanzees use more tools (Taï = 26 tools, Gombe = 22 tools and Mahale = 14 tools). In addition, Taï chimpanzees have been seen to make tools using six different techniques, whereas Gombe and Mahale chimpanzees used only three shared techniques.[2] So life in forests seems to favour tool use.

Food sources and insect diversity vary in different forests, reflected in the tool kits used by the local chimpanzee populations.

Forests in Central Africa contain a lot of insects together with an especially high density of bees. So, chimpanzees of the Goualougo Triangle in the Nouabele-Ndoki National Park, Congo, and in Loango National Park, Gabon, have been observed to make a large number of different tools for extracting honey from beehives. The chimpanzees of Goualougo Triangle have recently been found to use 28 tool types, that is, two more than Taï chimpanzees, and were also seen to make specific use of tool sets.[3] For example, to extract the honey from beehives located high up in trees, they used three different types of tools. These beehives are difficult to access as the bees close the entrance with wax that hardens, and the honey is found deep in large hollow branches with many chambers also closed by wax. To gain access to the honey, the Goualougo chimpanzees first use a pounding tool, a stout branch, to break open the entrance, then they use sticks as levers to open and widen an access point within the hive leading to the honey and larvae-filled combs, and finally they use smaller sticks to extract the honey from the hive. The chimpanzees in the rainforest of the Loango National Park were also seen to use equally well-developed tool sets for extracting honey from beehives.

The use of sticks by Goualougo chimpanzees is particularly impressive. They insert the stick vertically in the ground creating long and narrow tunnels which reach the subterranean chambers of termite nests. From the surface, a termite nest presents no obvious signs of its underground structure; thus, chimpanzees have to repeat this digging procedure until they feel a difference in resistance indicating that a chamber has been reached. They then remove the stick and insert a thinner straight fishing stick into the tunnel. They wait for the soldiers to cling to it, then withdraw it slowly and eat the termites one by one.

The use of tool sets seems to be a unique feature of forest chimpanzees. To the best of our knowledge, Goualougo chimpanzees use three types – that is, for honey gathering, fishing for underground termites as well as fishing for termites from high mounds – whereas Loango chimpanzees use tool kits suitable for both arboreal and underground honey extraction. Finally, Taï chimpanzees use a tool set for extracting almond remains from broken nut shells. Thus, these three forest chimpanzees all use tool sets for different purposes. So far, there have been no reports of this kind of behaviour in other chimpanzee populations.

The contrast with the three chimpanzee populations in Uganda that have been studied for over 10 years cannot be more dramatic. Tool use is very rare, almost absent: chimpanzees in Ngogo have been seen to use at most three types of tools, while those of Kanyawara in the same forest use only two types. The chimpanzees of the Budongo National Forest, which have been studied for over 15 years, were not seen to use any tools to forage for food. The deciduous forests in Uganda possess a rich variety of large trees with a wealth of insects, such as driver ants, termites, wood-boring bees, sweat bees and beetles, all of which are exploited with tools by other forest chimpanzees. So why do these chimpanzees not eat them and, as a consequence, not use tools to obtain them? Or, do they not use tools and, therefore, not eat them?

Despite Louis Leakey's misleading choice for his ground-breaking comparisons, the question remains as to what favoured the development of tool-using abilities in chimpanzees. If my explanation of the forest providing more opportunities for tool use in terms of increased access to rich insect resources applies, then we can understand why Gombe and Mahale chimpanzees use fewer tools. However, that does not explain the almost complete absence of tool use in forest-dwelling Ugandan chimpanzees. The prevailing deciduous forest type there is drier, with the majority of trees losing their leaves at the same time every year, compared with the mix of evergreen forest in Goualougo, Loango and Taï with an average of 1,800 millimetres of annual rainfall. For some insect species favoured by chimpanzees, such as ants and termites, humidity is very important and this requires not only abundant rainfall but also constant forest cover. So, it may well be that the Ugandan forests are poorer in insects and insects are not used by chimpanzees as a food resource. Whatever the reason, the scarce tool use by these chimpanzees remains intriguing.

The absence of support for Frodo when attacked by red colobus monkeys was surprising to my Taï chimpanzee eyes. Support, even when just vocal, would be provided in Taï whenever I saw chimpanzees facing the aggressive attacks by the colobus. Is my impression of less support in Gombe chimpanzees confirmed by more observations? Intriguingly, Gombe chimpanzees do not tend injured group members if they are not close kin. This observation was made by Jane Goodall and she noted particularly that individuals seemed to be actually repelled by the sight of injuries and turned away from them. Similarly, when Ntologi, the year-long alpha male in the Mahale chimpanzees, was badly injured during a fight with some challengers, no other chimpanzees offered him any help or cared for his injuries and he eventually died from them. Other males have been reported to suffer bad injuries during within-community fights and needed many months to heal. Again, no mention was made of any help or care from other group members. This is in sharp contrast to the continuous and extended care provided for up to several weeks that we observed between Taï chimpanzees, independent of kinship. As mentioned in Chapter 3, all injured individuals received a lot of support from other individuals and their injuries were licked and cleaned for weeks. Injuries healed very quickly in Taï, except for the few fatal wounds that we witnessed.

Support during intercommunity hostilities has never been reported from Gombe and Mahale and is rare in Ngogo chimpanzees. Dramatic descriptions have been documented of badly bitten chimpanzees trapped by neighbours, in auditory contact with their community, but nobody came in support. There are no reports of attacks being interrupted by the sudden appearance of a rescue team.[4] It is true that Gombe and Mahale chimpanzees can be more solitary than Taï chimpanzees and, therefore, surprised individuals might often be more isolated. Nevertheless, the distress screams accompanying such attacks might have alerted others in the vicinity and support could be provided. Both these differences are intriguing.

Both lower levels of sociality and altruism could be explained by the absence of leopard predation on chimpanzees at Gombe and Mahale. Leopard attacks happen regularly in the Taï forest and, therefore, the selective pressure to find a counter-strategy to deal with them is very important, indeed life saving. I suggest that it requires a certain frequency in the attacks by leopards to make the reciprocal dimension of altruism worthwhile.[5] Once altruistic behaviour is established, help could be provided in other contexts, such as during intercommunity aggressions. The relative rarity of life-threatening injuries resulting from violence within and between communities, in Gombe or Ngogo, might not be sufficient for the widespread use of altruistic and cooperative behaviours, compared with the high frequency of life-threatening injuries resulting from leopard attacks in Taï chimpanzees. If this suggestion is true, this could explain the greater levels of cooperation when hunting under the challenging conditions of the continuous canopy of a rainforest.

ANOTHER LOOK AT HUNTING IN CHIMPANZEES

Hunters need to be successful or they will stop hunting. There is no reason why we should expect cooperation to be the only means to achieve this: just look at all the solitary hunters, like leopards, tigers, etc. Cooperation is favoured under difficult conditions where individuals alone cannot solve the problem they face. Group hunting is seen regularly in different chimpanzee populations and this seems to be related to the number of males in a group. However, the way they organize themselves during the group hunt varies under the different conditions that chimpanzees encounter when they hunt.

One day in summer 2002, John Mitani and I followed the Ngogo chimpanzees of the Kibale National Park in Uganda. At the time I visited, this community was outstanding for having 24 adult and about 14 adolescent males. The sight of such a large number of males moving together in the forest is unforgettable. We were moving in a valley in the east of their territory and arrived during the heat of the day under a large group of red colobus monkeys assembled in three tall trees. Many of the chimpanzee males seemed attracted by the monkeys and moved around under their trees, scrutinizing the canopy. Some males started to climb up into the trees, which caused the monkeys to give alarm calls. To my surprise, given my experience with the Taï chimpanzees, many of the hunters sat down on the branches of the tree some 10 metres under the colobus monkeys, and just watched them and did not climb higher up to their level. At different times, one of the chimpanzees moved to a neighbouring tree where there were more colobus and was greeted by the threatening calls of the males. Some colobus males charged in his direction, but he waved a hand at them and they all quietened down. Over and over again, the chimpanzees would move between these three trees and the red colobus males would react by threatening them. To my Taï chimpanzee eyes, the situation was extremely puzzling as the monkeys, cornered in two directions in the trees, could easily be forced to jump for their lives, possibly falling into the hands of the chimpanzees waiting on the ground – rather easy prey, I thought. From my observations during previous visits to Gombe and Mahale, however, I knew that the chimpanzees can be quite respectful of threatening male monkeys and do not face them directly.

Taï chimpanzees often initiate an attack to make the male colobus counter-attack and use this occasion to try to capture them, and they do the same with the much larger black-and-white colobus males. In Gombe, on the other hand, I and many other observers have seen adult male chimpanzees chased away on the ground by one adult male monkey! Somehow, the unfavourable hunting conditions in the Taï forest have forced the chimpanzees to overcome their initial fear of adult monkeys that is seen in all other chimpanzee populations.[6]

Since Gombe and Ngogo chimpanzees are still fearful of adult monkeys, they must have developed ways either to selectively hunt only infants and avoid adults or, by their presence, provoke some disorganization in the monkey groups.

To return to the day that I followed the chimpanzees at Ngogo: as the tug-of-war between the monkeys and chimpanzees developed in the trees, the monkeys started to scatter and this was noticed by the efficient hunters among the males. Then, Miles dashed towards a somewhat isolated juvenile on a lower branch of the tree. As he did so, two male monkeys chased him, but in the process isolated another juvenile to the east. Miles rushed behind this one, again pursued by the two males. The juvenile had to jump lower to escape, while a second chimpanzee, seeing the opening provided by the two males chasing Miles, rushed toward this second juvenile monkey. A kind of a chaos developed, with chimpanzees dashing in different directions towards monkeys which were less protected and several male monkeys efficiently trying to protect their group members by chasing and jumping on chimpanzees. Miles had to rush to the ground but caught his monkey, while another monkey fell to the ground after a pursuit in the tree. I saw an adolescent male chimpanzee holding a young monkey but he was immediately attacked by two monkeys and had to run away, releasing the monkey in his panic. One monkey fell unnoticed near my feet and climbed back to hide in the branches. Another one fell to the ground and was chased by a male chimpanzee. Another chimpanzee on the ground alerted by the bark of the hunter joined in the chase and both disappeared in the undergrowth behind the monkey. The pandemonium stopped progressively as five infant and juvenile monkeys were caught by different chimpanzees and the main hunters started to eat, with some meat beggars, including mothers that seemed to appear from nowhere, clustering around the meat owners. During the entire hunt, the main group of monkeys did not move from their tall tree refuge and some chimpanzees also stayed just below them. The Ngogo hunters were simply too numerous and the red colobus males could not effectively repel them. There is also a clear preference for young prey in Gombe and Mahale chimpanzees in contrast to the Taï chimpanzees, who target adult monkeys, which make up half of their captures.

The Ngogo chimpanzees are impressive hunters with an exceptional high rate of success and they very often achieve multiple captures during a same hunt. In this sense, they are clearly more successful than the Taï chimpanzees as they make the best use of an interrupted forest that permits cornering of the prey, a strategy simply not

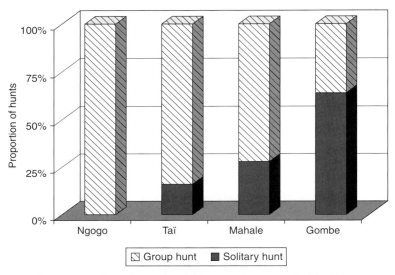

Figure 6.1   Group hunts, in which more than one individual hunts at the same time on the same group of prey, and solitary hunts compared between four chimpanzee populations, when preying on the same species of prey, i.e. the red colobus monkeys. These four populations are those for which sufficient detailed observations have been made to enable such a comparison.

available to the Taï hunters, which do not need to team up and cooperate. As can be seen from Figure 6.1, group hunts are common at Ngogo, more so than in Taï, which may be explained by the relatively larger number of males there. In other populations, like Gombe and Mahale, which have a similar number of males as in Taï, group hunts are clearly less frequent. The main reason may be because in these habitats the irregular tree cover makes hunting for monkeys very successful even for a solitary chimpanzee, as cornering them is facilitated by the interrupted canopy: lone hunters are five times more successful at Gombe compared with Taï, where the continuous canopy of the rainforest makes solitary hunting almost impossible. The challenge there is to hunt in groups or hunting ceases altogether.

As shown by the Gombe and Ngogo chimpanzees, group hunting does not necessarily require cooperation to be successful, as each participant may try his luck more or less independently of the others, whereas hunting within a continuous canopy requires a precise coordination of the activities of different hunters, as the monkeys could run away in different directions in the trees and quickly vanish. Therefore,

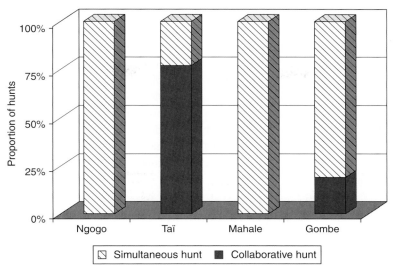

Figure 6.2   Frequency of collaborative group hunts, where
complementary hunting roles are performed by the different hunters,
compared with simultaneous hunts, where several hunters simply hunt
at the same time on the same group of prey, compared for four
chimpanzee populations.

as seen in Figure 6.2, Taï chimpanzees are the only population who
systematically organize group hunts with different, complementary
roles. As the Gombe chimpanzees have also been seen to collaborate,
it is not a question of ability but rather a question of necessity: in the
Taï forest, collaborative hunts are almost the rule and without this
collaboration, hunting would be much rarer.

Thus, here again, as in defence against predators leading to cooperation, it is not 'coming out of the forest' that favours cooperation:
it is the solving of the difficulties that chimpanzees encounter as a result
of living in forest. For all of the populations studied, monkeys are the
most abundant and easiest prey to locate and all chimpanzees show a
preference for red colobus monkeys. However, ease of hunting depends
upon forest structure and it is the chimpanzees' desire for meat that
forces them to cooperate to overcome difficulties. A similar process has
been observed in lions; in the Serengeti National Park where prey is
common and easily visible, the lions hunt mainly alone without elaborate cooperation, while in Botswana, where there are fewer prey and
better visibility, lions have been seen to adopt highly cooperative hunting strategies reminiscent of those seen in Taï chimpanzees.

The proposed 'out-of-savannah' model leading to the evolution of specific human abilities is still favoured by many anthropologists. What is the effect of forested habitat on these abilities in chimpanzees? The evidence from tool use and tool making contradicts this model; forest chimpanzees show more varied tool use and tool-making techniques than those living in more open habitats. It is the same for group hunting and collaboration, where more cooperation is observed in the forest in contradiction to the savannah model predictions. Furthermore, in the case of hunting, different factors make a hunt difficult and the solution to overcoming them is cooperation, as seen in chimpanzees and lions.

FEMALES AS A HIGHLY SOCIAL SEX

I proposed that high predation pressure within a rich habitat like the rainforest produces the conditions for both sexes to develop a cooperative social life similar to that seen in Taï chimpanzees. Leopards are ubiquitous predators in most rainforests of the African continent where their density is known to be higher than in open forests, woodlands and savannahs. Sadly, although most chimpanzees still alive in Africa are inhabitants of tropical forest, we have few studies on forest chimpanzees to test this idea systematically.

Leopards are absent from many of the East African sites. The Gombe National Park in Tanzania has a very small leopard population that almost became extinct 30 years ago, and no interactions between chimpanzees and leopards have been documented.[7] In the Mahale National Park, some leopards have been observed but no instance of predation on chimpanzees has been reported.[8] Four chimpanzees disappeared when lions visited the chimpanzee territory in 1989. In Uganda, leopards and other large predators seem to be absent from the northern forested parts of the Kibale National Park where the Ngogo and Kanyawara chimpanzees live, and they are also absent from the Budongo forest. The absence of predation pressure seems to directly affect the sociality of chimpanzees. Predation pressure by leopards seems to select for large parties all year round, with the females having a similar size range to the males (see Table 6.1).

A larger party size for long periods of time because of predation results in greater feeding competition between members of the community. Since food is very important to females as they spend most of their time either being pregnant or breastfeeding, it should not come as a surprise that females have developed special social strategies

Table 6.1   *Some aspects of the sociality of chimpanzees compared for four chimpanzee populations*

Since females have rarely been studied in much detail in some of the long-term populations, only a few of them can be compared.

	Populations of chimpanzees			
	Taï	Gombe	Mahale	Kanyawara
Habitat				
Forest type	Rainforest	Woodland	Woodland	Deciduous
Leopard presence	++	−	+	−
Sociality				
Party size	10.0	5.6	6.1	5.1
Male–female parties (%)	61	30	52	52
Home range overlaps (%)	90	56	75	65
Intergroup conflict support	Yes	No	No	No
Female intergroup participation	Yes	No	No	No
Male-female relations				
Grooming proportion (%)	34	49	21	28
Association time (%)	12	7	5	10
Coalition	Yes	No	No	No
Male mating initiation success (%)	74	76	78	−
Female mating initiation success (%)	77	40	17	−
Female-female relations				
Female solitary time (%)	1.8	65	11	23
Association time (%)	11	5	5	8
Grooming proportion (%)	22	13	−	0
Peripheral individual	No	Yes	?Yes	?Yes
Friendship	Yes	No	Few	No
Coalition	Yes	No	Rare	No
Female choice	Yes	No	?	No

to cope with the greater feeding competition. Table 6.1 shows that female Taï chimpanzees have larger home ranges than those in Gombe, Mahale and Kanyawara, form coalitions with other females as well as with males, have female friendships and groom both males and females that are not part of their immediate circle of friends, which has the distinct advantage of reducing the level of aggression between females. This seems an effective mechanism to control aggression in a highly competitive social setting.

This higher level of sociality among females seems to produce some additional developments in the sexual domain, whereby females gain more control and have been seen to enforce their choice of male partners and impose their choice against stronger males. A good measure of female choice is the rate of success of matings initiated by females. Studies in Gombe give the impression of females dominated by males and males being able to impose their will on females: the females' mating initiations are much less successful than those of males (see Table 6.1). In Mahale, the situation is similar, with almost all the females' attempts at courtship with adult males failing when a higher-ranking male is nearby, such that only 17% of female solicitations are successful.[9] In Kanyawara chimpanzees, the males are able to coerce females through aggression to mate with them and, therefore, impose their will. We observed exactly the opposite in Taï chimpanzees.[10]

The females' control over their reproduction seems to be variable in wild chimpanzees and clearly Taï females are distinct in having much more control than others. This shows how a relatively small difference in habitat, namely the presence of predators, can lead to a cascade of social changes in the prey species. Such a domino effect results from the interaction with pre-existing social abilities; the greater sociality of females allows them to form coalitions because of a pre-existing ability to form triadic interactions in the social domain. Female coalitions in turn permit them to impose their interests and this is in a species where traditionally the males are always dominant. This domino effect emphasizes how difficult it is to predict precisely the consequences of changes in social domains as theoretical predictions tend to emphasize the asocial dimension of females, the dispersing sex.

Why has female control over reproduction been lost in some chimpanzee populations? As the cases of the lions and langurs have already shown us (see Chapter 2), even if female control over reproduction makes a lot of sense to her and internal fertilization allows her potentially to make the last choice, the superior physical strength of the males can force females into situations where they can only opt for the least bad solution. Female lions and langurs have been seen to fake sexual receptivity to lure males into believing that they do not need to kill their young, but they are only able to do that if their offspring is already quite old or when they are still pregnant. In addition, they were not seen to be able to prevent males from killing their young. The conflict of interests between the sexes is obviously seen in these two species, but it is always present even if it expresses itself in less dramatic ways in most species.

Without predation pressure, increased sociality expresses itself in females mainly by greater competition for food. In East African populations, females seem to avoid this by becoming less social when food availability decreases. They return in large parties once food availability increases. Thus, only in very rich habitats is greater sociality a solution for females. In Gombe National Park, where dense forest covers only 4% of the park and grassland covers 33%, competition for food seems too great for females and they are much less social than Taï chimpanzees. Even so, high-ranking females are able to secure better ranges within the group territory and thereby achieve higher levels of survival and quicker maturation of their young. Female competition in such situations could be exacerbated and lead to the killing of infants of recent immigrants or low-ranking females by resident dominant females in Gombe and Budongo.[11] With high predation pressure, females are forced together and profit largely in terms of survival. Therefore, they adopt social strategies to adapt to the greater feeding competition. Taï females have female friends that support one another in contests, help to repair relationships after conflicts and have much fewer conflicts with one another.

The conflict of interests between the sexes is weakened as females have become social partners of males. Females help males to control access to meat during hunts. Similarly, females have been seen to support males for increased social dominance. Both aspects most probably profit some of the dominant males' reproductive success. In other words, some males profit directly from the greater control females exercise over their reproduction.

WARFARE PERSISTS IN ALL CHIMPANZEES

Long-term studies of different chimpanzee populations have revealed that warfare is present in all of them and that it can lead to the extinction of whole communities. Jane Goodall was the first to describe the repeated attacks of a large community on a smaller neighbouring one, the patrolling in the weaker territory and the violent fatal attacks against isolated individuals. She shocked animal lovers by describing the killing of a stranger by groups of males, and provided vivid descriptions of the violent concentration of strikes and bites on an isolated individual and gruesome behaviour patterns.

At first, the descriptions of warfare in the Gombe populations were greeted with scepticism, as if seeming to imply that the observation of warfare in wild chimpanzees gave some moral justification to humans for going to war! Consequently, it was dismissed as a result of the artificial provision of food that occurred in Gombe in the early days of the study.[12] However, the observations in the Taï forest and then in other chimpanzee communities, where none of the individuals were provided with food or even before they were habituated to humans, removed any doubt that such behaviour is natural and part of the repertoire of all known populations. The presence of warfare does not mean that it is carried out with the same frequency. The killing of strangers especially seems to occur at different intensities in different populations. For example, at Taï, in 29 years of observations, it was never seen in the North Group, although there was regular patrolling and intercommunity encounters occurred once a month on average. At the other extreme, eight intercommunity killings were observed in 7 years in Ngogo chimpanzees. As discussed in the previous chapter, in Taï chimpanzees, the presence of two new and ambitious males in the South Group might have been the cause of the three killings we saw there recently. In Ngogo, the especially large number of males present in this community, about 24 adult males, might explain the high level of fatal aggression. However, more differences were observed.

One key difference in the conduct of war is the frequency of support provided in Taï chimpanzees, support that was never or rarely seen in other chimpanzee populations. Detailed observations of attacks on single strangers have been reported in Gombe chimpanzees when other strangers were in auditory distance during the time that the noisy attack developed, but support never came. Similarly, in dramatic instances of attacks on their neighbours in the Ngogo chimpanzees, the neighbours were attacked and in the panic isolated strangers then became the target of the attacking party, but clear support has not yet been reported.[13] This is in dramatic contrast with the truly heroic support provided so regularly in Taï chimpanzees.

This absence of support could be paralleled with the absence of care for injured individuals in Gombe chimpanzees. This again is in striking contrast to the systematic tending of wounds by adults witnessed in Taï chimpanzees. Severe injuries resulting from intercommunity attacks or from leopards are much more frequent in Taï chimpanzees and can affect all individuals. This seems to have led to generalized altruistic care for others. Its restriction to close kin in Gombe chimpanzees remains a puzzle, but concurs with the notion that altruism will only appear once the reward to the individual exceeds a certain level.

A second difference in the conduct of war is the participation of females. Females in Gombe, Mahale or Ngogo chimpanzees have never been observed to take an active role in intercommunity conflicts. They sadly are victims and only ever passive. In Gombe, three quarters of adult females encountered by neighbouring males were subject to severe attacks and some were suspected to have died as a result. In Ngogo chimpanzees, a similar impression of fierce aggression against stranger females has been reported with five observed attacks in 7 years and possibly five more leading to infanticide. Such attacks on females are extremely puzzling as they do not seem to favour mate access for the males. In Gombe, long-term data showed that 6 to 11 out of 17 females appeared to have emigrated as adults.[14] As in gorillas, a sense of insecurity within a group might lead females to leave. This was not the case in Taï. So it could be that increased violence towards females leads to higher levels of transfer of adult females between communities, while greater tolerance towards females and their offspring leads to temporary visits of females between communities. Two different strategies might exist in chimpanzees resulting from different levels of aggression towards females.

Intriguingly, xenophobia seems to be more sex specific against males in Taï chimpanzees, while it is also directed against females in Gombe and Ngogo chimpanzees. By being less social and sometimes having contact with more than one community, females can cause paternity uncertainty in males, and aggression towards the females followed by infanticide could result. It has been suggested that within-community infanticide in Mahale chimpanzees was carried out against females that mated too frequently with low-ranking males. With greater community fidelity in Taï females, xenophobia seems to be directed against rival males and females are viewed only as potential sexual partners. We saw from Nadesh's transfers that resident females were aggressive towards her but not the males (see Chapter 2). The absence of aggression in Taï males towards infants of stranger females is puzzling as the greater community solidarity should mean greater certainty over paternity. However, we saw that females in Taï often voluntarily visited the neighbours and sparing the life of the infants might have some later benefit.

Thus, warfare in chimpanzees, which developed within a social system with high within-community solidarity producing xenophobia that makes possible the coordinated killing of outsiders, persists in social settings with lower within-community solidarity. In other words, even if there is no predation pressure, chimpanzees continue to go to war to gain sexual opportunities, but the level of sociality decreases when it comes to cooperation during hunting and support for individuals in difficult situations.

THE BONOBO, A FURTHER COUSIN, BUT WHAT AN EVOLUTIONARY PUZZLE!

In the deep forest of the Democratic Republic of Congo lives another of our cousins, equally closely related to us as the chimpanzee but much less known. The bonobo, originally called the 'pygmy chimpanzee', was identified as a distinct species first in 1912. The bonobo gained fame for being a very sexual ape, where homosexual behaviour especially between females is very developed, with genital–genital rubbing being performed whenever social excitement is arising. Its Latin name, *Pan paniscus*, reflects its close relatedness with the chimpanzee and genetic studies suggest they separated less than 1 million years ago. From this close relationship, they should be the ideal test case to see if my ideas are correct. The situation is, however, complicated for two reasons. First, they have been studied only for a relatively short period of time and there are only two detailed long-term studies where identified individuals are followed.[15] Thus, our knowledge of the bonobos remains very fragmentary and short-sighted. The second is that on many levels bonobos are very different from the chimpanzees, and this has been a constant challenge for the evolutionary scenario as they simply do not 'fit'. For example, bonobos do not use tools, have not been observed to hunt, seem to have non-aggressive interactions between groups, with females that seem to dominate all the males in some situations, and they seem to live in large, very stable cohesive groups.

Female dominance and the presence of large groups in bonobos would agree well with my expectations that predation pressure in the rainforest brings the individuals together and therefore females may gain more power in the social domain. More studies on this species would certainly be welcome to confirm this fact. The absence of violence in intergroup contacts is not compatible with my views and if confirmed would be actually a major challenge to them.[16] This situation is not unusual with bonobos as the absence of tool use and hunting has been impossible to reconcile with the proposed evolutionary scenario of both behaviours. However, a new study on bonobos in the Salonga National Park is confirming the presence of hunting in bonobos, so that we have to accept in bonobos, even more so than in chimpanzees, the fact that we still need to learn more about them before being able to draw any firm conclusions.

LESSONS FOR HUMAN EVOLUTION

The out-of-savannah model postulated a specific effect of the environment on the evolution of so-called human behavioural patterns. However, comparison of different chimpanzee populations has continuously revealed that this central tenet of the model is not supported. Rather than supporting a simple ecological explanation, chimpanzee studies have shown that specific features of the environment can lead to the evolution of specific aspects of behaviour and that these are not related to broad habitat types.

The most important aspect of such a comparison between chimpanzee populations is that *each* population faces its own specific set of environmental and social factors, and the solutions found by the individuals within each population are different. The tendency to generalize from one population or setting to the whole species is thus bound to be wrong and unwarranted given the adaptability of the species. Two factors have contributed to making chimpanzees so flexible. First, chimpanzees have proven to be one of the most adaptable primate species on this planet: chimpanzees are found in the deepest rainforests as well as in very dry savannah regions provided there are some gallery forests. The more diverse ecological conditions they encounter, the more different solutions will be developed to survive and reproduce. Second, chimpanzees have extraordinarily elaborate learning abilities that allow them to react specifically to the different challenges they face.[17] In the scientific world, these abilities are not always sufficiently understood or taken seriously. This has led some psychological studies on captive individuals to underestimate completely the impact that living conditions have on the development of cognitive abilities.

Population specificity is one of the most important characteristics of chimpanzees and should be considered whenever species are compared. No single population is representative of the species, but what is representative is the flexibility in different abilities or behavioural patterns. All populations, therefore, are fascinating as they contribute to the richness of the species and allow us to gain a real insight into the species and its abilities. The fact that tool use is almost absent, hunting so successful and males so numerous in Ngogo chimpanzees does not mean that 'The Chimpanzee' lives only in large groups, hunts often and is a poor tool user. Similarly, the fact that tools are used almost daily, hunting is cooperative and altruism is present in Taï chimpanzees does not mean that 'The Chimpanzee' is a constant tool user with cooperation and altruism omnipresent. Chimpanzees, like all living beings on the planet, do what is possible with the genetic baggage they have inherited to solve the challenges they face on a regular basis. If there are no challenges, then the abilities to solve them will not be stimulated, but if, on the other hand, they are present, these abilities will flourish. The conclusion we can draw from the above comparison is that 'The Chimpanzee' is a very flexible species that will use either many or a very few tools, hunt often or rarely, live in large or small communities and demonstrate a lot or a little cooperation, and the females will have varying levels of control over their reproduction.

And more flexibility is bound to come! Studies on unknown chimpanzee populations, like those in Goualougo or Loango, will provide new insights into aspects that we may not yet have thought to be adaptable in chimpanzees, and new studies on females in Ngogo will complement our understanding of how flexible female sociality can be in this species. It is difficult for humans to be patient,[18] but we have been wrong so often in our conclusions on what chimpanzees can and cannot do, that it might be sensible to be more patient. In the early 1960s, chimpanzees were still thought to be unable to use tools and hunt for meat. In the early 1980s, it was claimed that chimpanzees were unable to cooperate and lacked culture. In the early 1990s, it was declared that chimpanzees were unable to feel empathy and imitate others. In the early 2000s, chimpanzees were said to be unable to have joint goals and sustain traditions across generations. The list could go on, but my point is simply that we seem unable to refrain from drawing strong conclusions from the major differences between humans and chimpanzees, even though we know that we do not know everything and that eventually new data will contradict these hasty conclusions. It seems much more fruitful to concentrate on what is known about chimpanzees and accept flexibility as one of their main traits.

It is fascinating that chimpanzees are so flexible in aspects directly related to ecological challenges, such as tool use, as well as in domains more remotely under the influence of a simple ecological factor, such as female sociality. From comparisons made on more simple organisms like insects and birds, we tend to think that animals have limited adaptability in social and reproductive domains. Owing to both the increased physiological flexibility and increased learning

ability of the primate family, chimpanzees and humans gained a decisive advantage in adapting to very different living conditions. This is
especially striking when looking at gorillas and orang-utans; these are
our second closest living relatives, but both are restricted to a very
narrow ecological window, that is, dense primary forests.

How did chimpanzees and humans manage to enlarge this narrow
window? Two behavioural patterns are unique to both: tool use and
hunting. I proposed that it is this combination that is special and not
each ability in itself.[19] Tool use has been observed in many animal
species, including birds, otters and monkeys, and hunting is the rule
for all carnivores. Through hunting in the forest, chimpanzees have
learned to anticipate the relationship between external cause and effect:
for example, how prey will react to the approach of a second hunter. An
improved ability to predict the relationships between external events
has allowed them to use increasingly complex tools as well as to understand the benefits of cooperation and altruism. This, in turn, made it
easier for them to invade new types of ecological conditions, where both
tools and cooperation made the exploitation of food resources more
profitable.

The behavioural adaptability of chimpanzees is the one major
discovery of the last 15 years. Chimpanzees' sociality is finely tuned to
the challenges they face, and when facing high predation pressure,
cooperation and altruism become key elements of their social life. The
chimpanzees' technology developed to increase access to food and
security in the daily challenges they face, and when necessary, they
use weapons to chase and attack leopards as witnessed in Taï forest,
and use complex tool kits to access resources of high nutritional value,
such as the honey in Goualougo forest. Humans too showed different
levels of technology, with, for example, aboriginal Tasmanians having
tool sets very similar to those seen in chimpanzees while many modern
western societies are surrounded by human-made tools and artefacts.
In both species, flexibility is the key adaptation for coping with different living conditions. It is through the daily dealing with challenges
that humans' abilities develop. When deprived of such challenges, then
these abilities are not developing.[20] So, any fair comparison between
the two species should be made on populations facing similar challenges: it is to be expected that humans and chimpanzees living in
forests will show more similarities than free humans compared with
captive chimpanzees.

There is so much more to discover! New studies on chimpanzees will reveal many new aspects about behavioural flexibility in

chimpanzees and how varied their sociality can be. Similar studies on our ancestors are in constant progress, with unknown ancestors being discovered almost every year[21] and new information comes to light about the ecological conditions they faced. All this stresses the need to remain cautious about delivering definite answers to the question 'What makes us human?'.

We, humans, are fascinated by ourselves and eager to learn about what makes us so special. What is it really that distinguishes us from the chimpanzee – so close to us in many aspects? As our knowledge of our cousins in the forest increases, we gain a sharpened view at the chimpanzee inside us. In the next chapter, in an attempt to satisfy our impatience, I will suggest some aspects in our behaviour that I think are fascinating aspects where we might be uniquely human.

NOTES

1. Jane Goodall's reports (1963, 1964) were followed by many showing that tool use is performed by all chimpanzees wherever they live and for many different purposes. However, the impression of these first reports remains very strong and the Gombe chimpanzees are considered to be the 'termite fishers'. The publications on the Taï chimpanzees using various large wooden and stone tools somehow made them known as 'the nut-crackers' (Boesch and Boesch 1981, 1983, 1984).

2. Tool use has been documented in all known chimpanzee populations and even when the individuals could not be seen, the remains of the tools were visible (such as modified sticks or used hammers). Our first comparison of tool use and tool making (Boesch and Boesch 1990) was followed by others and all agree that Taï chimpanzees are among the most proficient tool-using populations compared with other habituated populations that rely on tools much less.

3. Crickette Sanz and David Morgan made these observations in the very isolated forest of northern Congo with the Goualougo chimpanzees (Sanz et al. 2004, Sanz and Morgan 2007). Josephine Head and colleagues made the observations about tool use in Loango chimpanzees (Boesch et al. in press b).

4. These two differences are puzzling and were greeted with some scepticism when I first mentioned them in 1993. However, new observations have been made since and other scientists have been looking for such behaviour patterns. At the time of writing, all observations concur that these differences might be real ones.

5. Trivers (1971), Axelrod and Hamilton (1981) and Axelrod and Dion (1988) have worked on reciprocal altruism, and they showed that its evolution requires a certain level of frequency in social interactions. Whenever the chance of encountering each other again is very low, there is no chance that reciprocal altruism will evolve. Therefore, group living and long lifespan, as in chimpanzees, are both favourable preconditions for the evolution of reciprocal altruism. In such groups, frequent predation attacks would then set the stage for the development of reciprocal altruism.

6. By observing the chimpanzees of Gombe and Mahale hunting using exactly the same methodology as in Taï, I could see that the attitude of the chimpanzees towards the monkeys was different and that each chimpanzee population adapted to the specific ecological conditions they faced and found a unique solution to their problems. For example, Gombe chimpanzees searched for parts of the colobus group with no adult males and would then hunt for infants or juveniles. Mahale chimpanzees, on the other hand, would spend long periods of time on the ground under the colobus group, sometimes displaying and waiting for the monkeys to finally flee and become disorganized.

7. Leopards might well have been part of the ecosystem in Gombe National Park before the deforestation progressed so far in the last 100 years. Nowadays, Gombe is a small island in a field landscape. As in Mahale, it could, however, have been a region with low leopard densities and leopard attacks against chimpanzees might have been very rare. It remains that once leopards do not represent a direct predation risk, chimpanzees will be able to adapt more freely to feeding competition: in a region limited in chimpanzee food, they will decrease their sociality so as to limit the cost of feeding competition. Gombe National Park has been suggested to be limited in chimpanzee food and this has been proposed to explain the small body size of this chimpanzee population (Morbeck and Zihlman 1989, Pusey *et al.* 2007).

8. Chimpanzees in Mahale have been seen on different occasions to react very aggressively against leopards to the point that they have been reported to attack a mother leopard within a cave, steal away her cubs and kill them (Hiraiwa-Hasegawa *et al.* 1986). Such a reaction suggests some aggressive interactions between the two species, although no further observation has been reported.

9. In Gombe chimpanzees, females have been shown to initiate some matings but most are initiated by males; more importantly, the male initiations are very successful while the females' much less so (Wallis 1997). In Mahale chimpanzees, the control exerted by the highest ranking males seems even stronger and females were seen to mate more frequently with the highest ranking males when they were more fertile. In addition, the alpha male is regularly able to interrupt the mating of other adult males so that the overall male mating success is much lower than in Gombe (Matsumoto-Oda 1999).

10. The potential for aggressive males to mate more is seen in all males and leads to an increase of stress in the females (Muller *et al.* 2007). A similar study in Taï females showed that aggression of males against females is rarer and that it does not lead to more mating success (Stumpf and Boesch 2006, unpublished data).

11. The killing of infants by the high-ranking female, Passion, in Gombe, was initially thought to be pathological but further signs of dominant females attempting infanticides led to a revision in the interpretation of infanticide as an expression of strong competition between females (Pusey *et al.* 1997). A similar interpretation has been proposed for the observation of infanticide performed by females in the Budongo chimpanzees (Townsend *et al.* 2007).

12. Margaret Power (1991) was the first to develop the idea that warfare in chimpanzees could not be natural and had to be a result of human intervention. At the beginning of the project, Jane Goodall used to give bananas to the chimpanzees when they visited her base camp. This was done for many years and she described how it led to an increase in aggression between the visitors. This was the main reason why she decreased dramatically the amount of bananas provided for each chimpanzee.

13. We obviously have to remain sceptical about whether support really is absent in Gombe or Ngogo chimpanzees as generally an absence is much harder to prove than a presence. With time, examples of support may be collected. However, 8 years ago, I published my first observations of regular support in intergroup fights (Boesch and Boesch-Achermann 2000) and had discussions with colleagues working in Gombe and Ngogo and all seemed to agree that such support is at most rare in these populations.

14. In Gombe chimpanzees, some females were observed only irregularly and could be absent for months and even years (Williams *et al.* 2002). This made their identification difficult and researchers could not be sure of their exact number. In contrast, female Taï chimpanzees were seen on a very regular basis once they had transferred into the community and they remained permanent members of their new community for the rest of their lives (except for Jessica, one female of the disintegrating Middle Group).

15. This special point has been argued as being the main reason to explain some of the unexpected differences we observe between the bonobos and the chimpanzees (Stanford 1998). Furthermore, the incomplete level of habituation of the individual bonobos makes it very difficult to evaluate the impact of humans and the real behaviour. In addition, studies done with captive bonobos have for quite some time portrayed a bonobo social life (de Waal 1989), that field workers were not able to confirm from the forest of Congo both in terms of sexual behaviour as well as the social position of the females (Furuichi 1987, Hohmann and Fruth 2000). This has raised a whole debate about what we know about the bonobos and patience is needed to obtain the first confirmation of these observations. The civil war in the Democratic Republic of Congo made research in this country impossible or especially difficult for some years.

16. One study of bonobos in Wamba was done with the help of heavy artificial provisioning with sugar cane (Kano 1992) and this is known to affect grouping patterns. In Gombe, the banana feeding of the early 1970s attracted large groups of chimpanzees to the feeding site and even the neighbouring groups from the southern regions. As long as bananas were provided the situation was relatively quiet, but became more violent once they stopped providing them (Goodall 1986). So we cannot be sure what the relation between groups is in bonobo without the effect of artificial provisioning. The new study of bonobos in the Salonga National Park promises to expand our knowledge of this species and has already shown that hunting is much more prevalent than we have previously thought (Surbeck and Hohmann 2008).

17. Learning abilities have been shown to increase as the size of the brain increases in both birds and primates (Lefebvre *et al.* 2002, Reader and Laland 2002). The development of the brain of chimpanzees is remarkably long compared with all other primates, except for humans, and allows them to continue learning new techniques into very old age: nut cracking is learned well after the age of 12 years and hunting techniques are still being learned by chimpanzees over 25 years old (Boesch and Boesch-Achermann 2000). So, adult chimpanzees, like adult humans, are still capable of learning and this results in an important improvement in the performance of individuals.

18. The first experimental study daring to directly compare them revealed that chimpanzees are more patient than humans! Does this come as a surprise? Chimpanzees were willing to wait up to 3 minutes before obtaining a food reward in an experiment, while humans in the same experiment waited only 40 seconds (Rosati *et al.* 2007). Luckily for us humans, monkey species

were willing to wait for a few seconds at most. In other situations, humans have been seen to wait for longer, but it is interesting to see that in a directly comparable situation, humans fare more poorly than chimpanzees. I wait impatiently for more such direct comparisons!

19. Boesch and Boesch-Achermann (2000) proposed an evolution model emphasizing the commonalities of humans and chimpanzees rather than concentrating on the differences that are still basically unknown. We suggested that a flexible social organization based on fission–fusion of community members along with flexible and systematic tool use and hunting are the common characteristics of our ancestors and that the forest was the setting under which this evolutionary process started. In other words, some of the classically described 'human abilities' were in reality shared with chimpanzees for a very long time.

20. An obvious simple example is that chimpanzees would not know how to crack nuts if they never had access to them (Boesch *et al.* 1994). How could complex social abilities develop without dependable social partners? Many studies have shown that captive chimpanzees cannot cooperate (Tomasello and Call 1997), until the quality of the social relationship between partners was considered and then they cooperated well (Melis *et al.* 2006). In humans as well, some abilities will develop only with practice and some complex social abilities develop faster in individuals who face more stimulation (Perner *et al.* 1994, Cutting and Dunn 2006, Boesch 2007).

21. A week before I wrote this sentence, the discovery of two new fossils of *Homo erectus* and *H. habilis* totally changed the scientific world's knowledge of the human past and it will have to be accepted that the latter lived much longer ago than we thought and was even a contemporary of *H. erectus* (Spoor *et al.* 2007), leading to a new question about who were the ancestors of *H. erectus* and when did they live.

7

When sex becomes destructive

Louis David, the famous French romantic painter, was the official artist of the emperor Napoleon. One of his most lavish grandiose paintings, exposed today in the Great Gallery of the Louvre in Paris, symbolized the essence of human wars in the classic episode of the Romans kidnapping the wives and daughters of the Sabins. The painting shows the fights between the Romans and the Sabins, who were taking revenge for the kidnapping of their women, at the dramatic moment where the women holding their babies interposed themselves between the swords of their husbands and the spears of their fathers! David's painting illustrates vividly my point of view about war, in that the Sabin women were exposing the intricacy of the situation between enemies that are at once kin and killers. Furthermore, isn't David suggesting that the competition of men gets out of control as women and babies are the primary victims? Make war to get love! Or is it love that makes war possible? I would like in this chapter to concentrate on the question of why in humans sex is at the base of war and why women seem to suffer the most from it.

In a review of 230 human tribal groups, Keeley (1996) found only eight that sometimes spared the lives of adult male captives, while in the majority, women captives were kept alive to be used as sexual partners and/or as a cheap workforce. Sex and women are commonly mentioned as the principal reasons for war: within 186 human societies, blood feuds around marriages between kin groups are reported to represent the majority of the cases of violence. Further, often the more exchanges there are, including marital, between human groups, the more conflicts and wars are likely to occur, as in New Guinea. In Native North Americans, war was proposed to have resulted in 93% of cases as revenge for previous killings, in 60% of cases in retaliation for the poaching of cattle and other breeding animals, and in 58% of cases because of the capture of women.[1] Revenge is regularly proposed as an explanation for war, but it is often difficult to discover the initial causes of conflicts as they are sometimes forgotten in the intervening years. In a sample of 42 foraging societies, war was strongly associated with conflicts over women unless other transferable resources were available. In addition, when resources were transferable, the wealth of men and polygamy in the society were associated, while this was not the case when resources were inalienable. As seen above, women are often subdued and the stealing of women is regularly observed, for example, by Aborigines, Inuits and the Yanomamö. Logically, farmers and sedentary hunter-gatherers have to defend their resources or face severe privation. When threatened it is probably better to attack first rather than wait to be the victim. Since land tenure determines reproductive success in these groups, war fought over land is simply an extension of sexual conflicts.

Love is what makes war possible! Porthos was one of the 'Three Musketeers'. In 1844, Alexandre Dumas portrayed him as a glorious and robust epicurean with a big generous heart. The Three Musketeers symbolize some of the finest virtues of 'what makes us human' – fearless heroism, intelligence and shrewdness combined with the ability to use lethal violence when necessary for a good cause, together with life-long friendship and support provided with unlimited unselfishness – 'one for all, and all for one'. Porthos took this to extremes when, after supporting and being supported by his friends in countless fights against the bad guys, and been cared for by his friends after suffering life-threatening injuries, he sacrificed his life when he prevented a huge stone from falling until his three friends were safe. In addition to the marvellous pathos of the story, *The Three Musketeers* has remained popular over the years as it is a vibrant evocation of the uniqueness of human nature, that is, team spirit, boundless cooperation and ultimate altruism. There are many similar sagas of human heroes, more or less famous, found throughout the world.

The same kind of conviction can be found in the scientific literature as, for example, in 2006 in the prestigious journal *Science*, where an article was published on the enigma of human nature: 'In every human society, people cooperate with many unrelated individuals; division of labour, trade and large-scale conflicts are common. The sick, hungry and disabled are cared for and social life is regulated by moral systems enforced by third-party sanctions. In contrast, in other primate species, cooperation is limited to relatives and small groups of reciprocators'.[2] Such a statement typifies the contrast that some would like to be perpetuated, that is, placing humans in a different and higher category from all other living animals on the planet. At the same time, however, this 'wonderful moral human animal' produces levels of violence that have destroyed large numbers in some populations. For example, in the Waorani of Ecuador, 60% of adults over five generations met a violent death. This constitutes the world record! More average, among the Yanomamö of Paraguay, 15% of adults died because of war, including 24% of men and some 7% of women.[3]

When following Porthos with Gia on his back rescuing the one-armed female Bamu from outnumbering aggressive males, I was totally flabbergasted by his bravado and style, but I never thought for a second that he was a human. Asking for a fair recognition of what chimpanzees are does not imply that we should diminish humans with their unique capacities. After all and simply stated, I am writing this book and not Brutus, the long-time alpha male of the Taï chimpanzee community. Conversely, Brutus, not I, was able to capture an adult colobus by hand 40 metres high up in the trees after several anticipations of the monkeys' escape possibilities. The descriptions and arguments I outline in this book aim to provide a fresh look at what we know about chimpanzees and some traditional human societies and attempt a reply to the question of what makes us human. Desmond Morris (1967) in his famous book *The Naked Ape* reminded us that 'Our climb to the top has been a get-rich-quick story, and, like all *nouveaux riches*, we are very sensitive about our background. We are also in constant danger of betraying it'. Humans tend to believe there is something special about and in them, something that elevates them above the biological rules. But we are, beside all our big technical achievements, still and largely a simple biological phenomenon. We have already touched on this in the previous chapters but I want here to address more specifically what could have become different during the human evolution that rendered us different from the chimpanzees and try to understand how this could have happened.

THE TYRANNY OF THE TESTES

War is a general state in humans. Detailed recent reviews reveal that almost all of human societies have engaged in war at least once a year, while only 5% are truly peaceful. These last societies are often groups living in extremely low densities and isolated from other groups (often they are recently defeated refugees). These reviews indicate that the frequency of war seems to be decreasing as society develops and becomes more technologically sophisticated: 70–90 simple bands, tribes and chiefdom societies went to war at least once every 5 years, while industrial societies went to war once per generation.

Jean-Jacques Rousseau in an outpouring of wishful idealism proposed that before the advent of agriculture, humans were pacifists and did not engage in warfare. He proposed the myth of the gentle primitive, as nature would have imposed an inhibition on the killing of conspecifics, for the good of the species, and that it is only recently that humans have become uniquely violent. On the other hand, the English philosopher Thomas Hobbes said that humans, like animals, have great reproductive potential and, therefore, quickly reach 'the carrying capacity' of the environment they inhabit, leading to high levels of competition between individuals. Such aggression would than regulate all social interactions between and within human groups.

How old is warfare in humans? It is not always easy to determine the cause of death from skeletal remains alone. The earliest known burials showing evidence of violent death are some 24,000 to 34,000 years old. In Grimaldi, Italy, a child was found with the point of a projectile in the spinal cord, while in France, a skull with cut-marks on the forehead has been found suggesting that the victim might have been scalped, while in Czechoslovakia, violent conflicts seem to have been common, with mass graves of men, women and infants.[4] More recently, a late Palaeolithic cemetery at Gebel Sahaba in Egyptian Nubia, some 12,000–14,000 years old, used by hunter-gatherers, shows warfare to have been common and brutal, as 41% of the 59 men, women and children buried there had the points of stone projectiles associated with their skeletons and all the children had these wounds, execution like, in the skull or neck. Similarly, in the Ofnet Cave in Germany, two caches of skulls, some 10,000 years old, comprising 34 men, women and children, most with multiple holes caused by stone axes, have been found, arranged like eggs in a basket. In Western Europe, fortifications around settlements of the first farmers became widespread in the Neolithic period (4,000 to 7,000 years ago depending on the region).

Why are humans so often at war? My proposition is that due to the slow development of modern humans and the longer investment required in their offspring, men face greater competition than male chimpanzees in finding female partners for reproduction, which results in a higher level of intergroup violence. Review data seem to support this proposition, as mortality rates due to warfare are about three times higher in human societies than in chimpanzees: on average, fewer than a third of adult men in farming societies died from warfare, compared with a tenth in chimpanzees.[5] In addition, the mere anticipation of warfare can lead human societies to incur additional costs, often underestimated and without equivalent in other species.

Women pay a massive price for this exacerbated male competition. In many human societies, infanticide has always been directed against girls and has led to considerable alterations in the numbers of infants. The average sex ratio at birth in humans is 105.5 males to 100 females. Remarkably, in a large sample of 112 traditional societies practising warfare, the sex ratio for youngsters 14 years of age or less is 128 boys to 100 girls. In other words, about 25% of the girls have died since birth. To illustrate how costly war is, in these same societies, the adult sex ratio returns to even numbers with around 96 men to 100 women. Warfare requires such a heavy cost for men that warring societies artificially increase the number of boys by killing girls after birth. It is enlightening that if war ceases in these societies for 5 to 25 years, the sex ratio for youngsters less than 14 years of age returns to 113 boys to 100 girls, and if war stops for more than 26 years, it returns to normal, namely 106 boys to 100 girls.[6] Sexual competition can become highly destructive!

The intriguing question is how could humans, the altruists and cooperators par excellence, become so aggressive and start to prize the killing of conspecifics? As in all other animal species, sex and reproduction are of paramount importance to humans. However, human demography, like that of other great apes, is characterized by a slow maturation rate with long interbirth intervals and a lengthy investment in each individual offspring. So, as with chimpanzees, all males face the challenge of how to cope with the fact that for extended intervals most females within their group are not fertile and the only way to increase opportunities for reproduction is to look for extragroup matings.

Now, if sex is so fundamental, why do priests in many human societies relinquish reproduction? Does this not prove that sex cannot be so important? This argument has always been advanced by philosophers in arguments with biologists when they start to discuss human nature. Such a disagreement has been around for centuries and not much progress has been made. Perhaps none should be expected! This is mainly a disagreement about whether humans are members of the animal kingdom or, rather, are superior beings the Almighty created in His image and, as such, not subject to the same rules as those regulating other life on earth. I do not have any desire to try to resolve these debates. Nevertheless, I think the answers are to be found in the subjects under study, that is, humans in the present case, and for this, the only solution is to go out and document what they are doing. I hope that from the previous chapters I have convinced the reader that, better than any philosophical debates, direct observations have contributed in a unique way to an understanding of our closest living relatives. Consequently, I suggest we rely on observations of humans rather than arguing with scientists, the religious and philosophers about what humans are 'supposed' to be. Fortunately, there is a wealth of direct observation from anthropologists who have spent many years sharing the lives of humans in different cultures and so have contributed a wealth of data on our own human nature; that is, they have done exactly what I, and my colleagues, did with chimpanzees!

INEQUITY IN HUMAN REPRODUCTION

For most of our past, our ancestors lived as traditional foragers. Agriculture began only about 8,000 years ago, but our direct ancestors, members of the species *Homo sapiens sapiens*, modern man, had been living on the planet for about 100,000 years. So our past was mainly that of traditional foragers, also called hunter-gatherers, who lived by hunting for meat and collecting fruits, tubers and leaves. Sadly, many human hunter-gatherer societies have disappeared and we can only study the few that have managed to survive.

Traditional foragers live in groups of only 48 individuals, surprisingly similar to that seen in chimpanzees. In addition, if polygamy is observed in about 84% of human societies, the number of men with more than one wife in a given group is limited; only 15% of men have more than one wife.[7] In the resource-poor Kalahari region of Namibia, only 5% of men in the !Kung Bushmen have two wives. However, some of these polygamous men can be extremely successful. Shinbone, the most successful Yanomamö man, had 43 children and his father had 143 grandchildren! Thus, in traditional foragers, polygamy is observed in most societies and is 141 times more frequent than its opposite, polyandry, where woman have more than one man. With an even sex ratio at birth, polygamy implies that some men have no access to women and this leads to very high levels of sexual competition between men.

Do men profit from such sexual competition? In many primate species, there is a clear and strong positive effect of a male's social rank on the number of offspring he will sire. Is this the same in humans? The data collected in many different societies overwhelmingly say 'yes'. Be it in traditional forager societies, in peasant farming communities or in modern western societies, men with a higher social status or more wealth or excelling at hunting have more wives, more sex and more surviving children.[8] For example, in the Ache of Paraguay, good hunters have more wives and their children survive better compared with bad hunters. In addition, good Ache hunters attract more extramarital mates, and father and raise more children than poorer hunters. In the Xavante agriculturalists in Brazil, 16 of the 37 males in a village had more than one wife, and the chief had five wives with 23 surviving offspring. In the Yanomamö of Venezuela, warriors have more wives and three times more offspring than non-warriors.

Elsewhere, good !Kung Bushmen hunters from Botswana father almost twice as many children as poor hunters and their chance of survival to 15 years of age is twice as high. The difference between the two classes of men is not due to a higher degree of polygamy, but to the fact that good hunters get married about 5 years earlier than poor hunters and their marriages are much more stable. In addition, wives of good hunters are harder working than wives of bad hunters, so that 80% of their children survive to adulthood, while only 67% of the offspring of poor hunters survive. Finally, good hunters have much larger exchange networks and live in their camps with proportionally more kin and stable kin groups than poor hunters. Since a child's future fertility increases the later he is born in a family with a large number of older siblings, this all combines to make good hunters and their families more successful. Similarly, the farmers of Kipsigi in Kenya with more land have more wives and more surviving offspring than poor farmers. The better lands compensate for the costs of having co-wives. In the Dogon in Mali, men with more wealth have more wives, and for each additional wife a man has two more surviving offspring.

Such reproductive advantages were also found in European farmers in the eighteenth and nineteenth centuries. For example, during 1824–1896 in Sweden, landowners were more likely to marry than non-landowners and to marry at a younger age and have more surviving children. Even in today's USA, men with a high income father more children and report more sex than men on low incomes. Furthermore, educated men have more biological children than educated women, while more intelligent men and women have less sex and fewer offspring compared with less intelligent men and women.[9] So it seems that whatever the socio-economic system under which we live, men have an unequal access to women, and those with a high status get more access and have more surviving children than lower status males.[10] This situation exacerbates the sexual competition between men and seems a good reason for them to go to war.

Human males have much to win from gaining access to more women. Since historical times, both higher status and wars were means to achieve this goal. This is reminiscent of what we have seen in chimpanzees. However, it goes one step farther in humans. Why and how did male competition become detrimental to the female? Female Taï chimpanzees, as we saw, are active partners in the social challenges of the males. They successfully manipulate males during their fertile phase, and can impose their choice of mate and dominate many males when competing for important food resources like meat. Male dimorphism is similar in chimpanzees and humans and so it is not the physical differences that explain why human females seem to have lost some of their power.

Just as in the chimpanzee, the large testis of humans is indicative of a long history of intense male sexual competition. What has changed is that, unlike chimpanzees, human females have become dependent upon the support of males to nurse and care for their offspring successfully. While female chimpanzees carry and breast-feed their offspring constantly for 4 to 5 years, they do that on their own within an environment where additional food sources are available so that the infants become independent by the time the next offspring is born. Two key innovations seem to have made the situation in humans very different – birth intervals and territory size.[11] First, early in the evolution of *Homo sapiens*, the intervals between births were dramatically shortened and, as a result, women could produce more children. Consequently, human children are weaned at a younger age than chimpanzees (2.8 years versus 4.5 years), notwithstanding the fact that humans mature more slowly (mean age of first birth is 19.5 years versus 13.3 years in chimpanzees). Further, contrary to what is sometimes said, chimpanzee and human infants are equally helpless at birth, with brain size only 10% smaller than that of the respective adult of each species, and rates of brain growth are similar, reaching adult size at about 3 years of age.[12]

Although chimpanzees have a comparable developmental profile to orang-utans they are able to wean their young about 2.5 years earlier, due to their more omnivorous diet. Conditions in the African rainforest mean that chimpanzees are able to wean infants relatively early as they can support themselves independently with solid food, unlike orang-utan mothers who are not able to do so in the more impoverished Indonesian forests.

The second innovation in the human line seems to be that all hunter-gatherer societies live in much larger territories than those of chimpanzees; home ranges for some 240 forager societies are reported to be about 175 square kilometres, which is about ten times larger than the average chimpanzee territory. The more a society relies on hunting then the larger the territory. Accordingly, the average population density of humans at 0.18 individuals per square kilometre is much lower than the chimpanzees, which is more than one individual per square kilometre. This is because humans have colonized many more different habitats than the forest-dwelling chimpanzees. The emergence of *Homo sapiens* is characterized by the occupation of new types of habitat. The newly conquered savannah, however, generally produced a lower biomass than the tropical rainforest, while at the same time supporting a greater biomass of large mammals. In other words, food was more difficult to find for the women gatherers, while hunting men were more productive.

Both innovations pose an insoluble challenge for women. Earlier weaning with slower developing children means mothers have to care for at least two to three children at the same time and when having to do so in a less productive environment, the survival of the children is compromised. So, increasing female fertility came at a cost. The only way to overcome this handicap was to enlist the help of others to provide for the additional dependent offspring. The obvious candidate to help was the father. In a study of 141 forager societies, it was reported that the male contribution to the family diet represents on average about 64% of the food, ranging from a low 25% to 100% in some societies.[13] This contribution of fathers had a direct positive effect on women's reproductive success by allowing earlier successful weaning of the children and better maintenance of ovarian functions. In addition, in the Ache and Hiwi of Paraguay as in the Hadza of Botswana, where detailed studies have been made, men's contributions increase when mothers are nursing young children and are, therefore, unable to invest much in searching for food. This adaptive provisioning by fathers has been shown to be essential.

But the size of the testis emphasizes that fathers are men! In addition to being important providers of food for the family, they continue to invest time and energy in competing with other males. In this context, it is interesting to note that female humans have developed the unique phenomenon of the menopause, whereby reproduction is interrupted well before the women are senescent. Post-reproductive women often complement the role of fathers and become important helpers in their own daughter's reproductive efforts.[14] The increased dependency of early weaned children makes additional help more important compared with other species. Thus, in order to increase the number of offspring, human mothers have had to become dependent upon the fathers and other family members in ways not seen in chimpanzees and other primates. Women have become vulnerable to many controls imposed by men who need to be confident about the paternity of offspring, which results in a loss of freedom for the woman.

In contrast to female Taï chimpanzees that are able to select their mate and sire of their offspring, the freedom of women in all human societies is limited in different ways by men. The most obvious and widespread way is by violence; historically and presently, throughout the world, the less certain men are about paternity, the more they decrease their investment in the child or worse, abuse the child, and more often divorce or even kill the mother.[15] It is not uncommon for husbands that doubt the fidelity of their wives to hit them; a nasty Yanomamö husband could hit her with the sharp edge of a machete or shoot a barbed arrow in her buttock or the leg. Another punishment is to hold a glowing piece of firewood against the body of the wife. Husband's punishments are adjusted with the perceived degree of the wife's fault, with the most drastic measures being reserved for infidelity when some husbands kill unfaithful wives. Similar reports can be found in a large number of human societies. The main point being that all indigenous legal codes address the issue of men's entitlements to sexual access to women and adultery is generally conceived as unauthorized sex with a married woman, in which the victim is the husband, who may be entitled to damages, violent revenge or to divorce. This double standard has been found in such different societies as in the Inca, Maya and Aztec, in Germanic tribes, in African legal codes, in Chinese and Japanese societies, and in all ancient Mediterranean states, such as Egyptians, Syrians, Romans and Spartans. Human infidelity is first known to be punished in 1810 in France and then Austria.[16] Sexual jealousy in humans is complicated by the fact that fathers invest parentally: bad enough to lose a fertilization to a rival, but having to invest in another's baby is worse!

Studies of wife killings from a variety of societies confirmed that the majority were due to suspected or actual female infidelity or decision to leave the marriage. It is important to see that such violence against women by men is not random in the sense that men take into account the fertility status of the women, so that pre-reproductive girls and post-menopausal women often enjoy much more freedom than reproductive women. Similarly, wife beating was more frequent in polygamous societies with a higher level of sexual competition.

In addition, many societies possess traditions and rules to control the sexual activities of women as soon as they marry. One of the most common traditions in many societies is the tolerance or even support of wife beating that can leave women badly hurt (such physically impairing assaults against females are almost never seen in chimpanzees). More institutionalized control ranges from menstrual huts as seen in the Dogon, the foot-binding tradition of Chinese women, the veiling of Moslem women, the complete imprisoning of women in the harem of rich men in different societies, to complete clitoridectomy – the surgical removal of most of the external genitalia – and infibulation – suturing to close the genital labia.

Polygamy, the predominant mating system in humans, is generally associated with coercion by men. In 186 forager societies, polygamous marriages were pre-arranged for women but not for men, suggesting that parents benefit by offering up their daughters to the most influential men. In addition, societies often at war have a higher proportion of men and women in polygamous marriages. Polygamy is an outcome of male competition and it has costs for women. As a rule, men who can afford two or more wives profit by having more children, while the individual woman suffers by having fewer surviving children than if she lived in a monogamous marriage. In an example from the avian world, the dunnock, a small passerine bird, has an interesting, and indeed remarkable, sex life. In addition to the typical monogamous pairs seen in many bird species, they also form polygamous groups with one male and two females, as well as polyandrous groups with one female and two males. In dunnocks, unlike humans, we have the distinct advantage of being able to confirm paternity by genetic testing and the respective reproductive benefit of the three pairing systems for both sexes can, therefore, be quantified.[17] For the males, there is no question that polygamy is the best solution. Polygamous males produce on average 7.6 offspring, whereas monogamous males produce 5.0 offspring. Polyandrous males are in the worst situation with a dominant male having only 3.7 offspring and other males only 3.0. For the females, the situation is exactly the opposite. Polygamy is the worst system as females produce only 3.8 offspring, but have 5.0 in monogamous unions and 6.7 in polyandrous unions.

The message is clear: males should prefer polygamy and females polyandry. For Dogon women, polygamous marriages have a further disadvantage as child mortality is about six times higher than in monogamous families. This is similar to that observed in the Kipsigis or Datoga of Kenya where women have more surviving children in monogamous unions than in polygamous ones, whereas men fare better in polygamous situations. So, as in the dunnock, polygamy is of clear advantage to men.

MAN THE HUNTED: ALTRUISM AMONG COMPETITORS

When Porthos heard the screams in the south he did not hesitate for long. He put his adoptive little girl on his back and ran off to support Bamu, the female who had been taken prisoner by the enemy of the South Group. He was risking his life, and with a baby on his back, was probably somewhat restricted. Despite this, however, he did not hesitate and his heroic move freed the female from her tormentors (see Chapter 3). Like Alexander Dumas's musketeers, the chimpanzee Porthos also provided selfless help, for both a baby that was not his offspring and an unrelated female of his group. The cost could have been high, but he did it nevertheless. A few minutes later, the defeated enemy captured and killed another member of Porthos' community. This duality of a high level of altruism and cooperation within a group and a considerable amount of violence between groups is strikingly similar between chimpanzees and humans. Could this have appeared for the same reason in both species?

I suggested earlier that such altruism and cooperation in chimpanzees resulted from the need to limit the costs of high predation pressure. Humans too suffered predation pressure from large predators. Our ancestors, bipedal and often armed hunters, throughout the past were confronted by fearsome predators: sabre-toothed tigers, giant cheetahs, lions in Africa, Europe and Asia, the cave, brown, polar, moon and black bears, hyenas, lion cougars, the *Smilodon* cats with their 14-centimetre canines in North America, leopards in Africa and Asia, jaguars in America, wolves and giant and spotted hyenas, to name just a few! All these predators shared their habitats with past and present-day humans and all have been shown to kill and eat humans.[18] Numerous accounts can be found in the scientific and other literature of the attacks that humans suffered from them even in the very recent past. To give just one example, as shown in the 1996 Hollywood film *The Ghost and the Darkness* with Michael Douglas as the hero, the railroad construction of the Uganda Railways in 1896 was stalled for many months in the Tsavo region because of lions attacking the workers – about 135 of them were eaten in 9 months! Marks on the bones and skulls of human remains are a testimony to the fact that humans were hunted by predators for millennia. Such predation against humans has been part of all our past, and was already noted with one of our earliest ancestors, the famous Taung Baby, an *Australopithecus africanus*, found by Raymond Dart in 1925 in South Africa that had two distinct tooth marks from a leopard on its forehead. That child was found in a cave where large cats had accumulated remains from early humans that they trapped, killed and most likely ate.[19]

Love is what makes war possible. A similar process to the one described in chimpanzees could have forced humans facing such deadly attacks to become more supportive towards one another, and this led directly to increased individual survival and survival of group members. This supportive behaviour combined with sexual competition led to the dualism of strong within-group solidarity and external xenophobia. Within-group solidarity in humans has been shown to correlate with hostility towards neighbouring groups. Many human societies, as in the Yanomamö of Venezuela, also called 'the fierce people', is based upon an unquestioning solidarity based on kinship, co-residents and friends coupled with a philosophy of revenge and violence. Social organization depends on a principle of solidarity that assumes the involvement of all in the misfortunes of one. Therefore, revenge is seen as essentially positive resulting from their obligation to solidarity and justice for kin and allies. The code of revenge is bound directly to an ethic of care and solidarity. Similar accounts have been documented for many different traditional societies, such as the Ache, Inuit, Nuers from Sudan or the Aborigines.

However, let's not forget the large testes of human men. Humanity's dualism means that within-group solidarity is tempered by the actual situations faced. In other words, we should not expect humans to *always* cooperate and be altruistic, but to adapt their behaviour to the competitive context in which they live. We saw that chimpanzees not facing high predation pressure were less cooperative and this is also to be expected in humans. The ability to cooperate is not a genetic 'yes/no' answer, contrary to some broad assertions that have been made when contrasting humans with chimpanzees.[20] This has been spectacularly illustrated in a study based on the 'ultimate game' that was played in 15 different human cultures, ranging from traditional foragers to farmers and university students.[21] The ultimate game requires an individual to offer a share of his money, equivalent to one month's salary, to another individual who can refuse if he feels the offer is not acceptable. If he refuses, neither keeps the money, if he accepts, the money is shared between the two according to the offer. Western university students behave according to the human altruistic model and tend to offer 40–50% of the money and refuse offers below 30% of the sum available, as they feel the offer is too low. Economists were surprised by the results as they expected humans to offer no money at all following a purely selfish strategy, forgetting that all humans grow up in societies with clear social rules.

Not all humans are equally selfless. When comparing the results from 15 human societies, it is fascinating that offers and refusals could vary substantially across societies. In seven societies, including the Ache of Paraguay, average offers were similar to those of the university students but they almost never rejected offers, no matter how low they were. In real life, Ache hunters routinely share all the meat they bring back to their camp and do so spontaneously in ways that are intended not to make the hunter appear generous, such as leaving the meat on the ground at the entrance to the village. On the other hand, members of three societies, including the Hadza from Botswana, offered much less than the university students and very often nothing at all. In real life, Hadza hunters share meat extensively, but they do so grudgingly and many look for opportunities to avoid sharing by hiding the meat, knowing they would be punished for failing in their social obligations. Finally, some populations actually rejected low offers but also refused very high offers! In those societies, accepting an offer brings an obligation to reciprocate and, therefore, high offers were refused. Interestingly, the tendency to reject offers is higher in populations that would offer more in a game where there is no punishment. In other words, groups with a high level of altruism reject low offers more often.

Modern *Homo sapiens* has always been a victim of many predators. In a process similar to the chimpanzee, he developed high within-group solidarity because, as long as efficient weapons were not available, systematic and long-term tending of injured group members and unfailing support for group members attacked by predators were the only possible ways to limit the negative impacts of such attacks. Such adaptable responses towards helping others only evolve in animal species that are able to appreciate that the needs of other individuals can be alleviated by the actions of third parties. So empathy is a key element in the emergence of such responses to predation. Predation is faced by many animal species but very few have been observed to support one another in such situations.

HUMANS AS THE THIRD CHIMPANZEE

In the past, humans and chimpanzees shared many attributes that have helped shape them. Both were long-lived primates with slowly maturing offspring requiring a long-term investment on the part of the female and lived in large social groups facing predation pressure. In large-brained animals, these circumstances select for high levels of within-group solidarity as well as strong sexual competition between the males. In both species, males show some of the highest forms of altruism and cooperation towards group members whether it be in the context of defence against predation, or in adopting and supporting weak group members, sharing high quality food like meat with non-hunters, or adopting complex cooperative teamwork in hunting. In both species, intergroup aggression is high and the drive to find more sexual opportunities and guarantee the successful development of offspring seems to be at the origin of this violence.

Similarities between humans and chimpanzees do not stop there. Tool use is another major shared skill; all populations of both species have been seen to use tools and adapt them for different purposes. Both species have gained access to new niches, enriching their food resources not accessible to species not using tools: hard-shelled fruits, underground tubers, water, insects or honey have, because of tool use, become regular dietary items of chimpanzees and humans.[22]

All these similarities taken together set humans and chimpanzees apart from all other living species. Some animal species may possess one or other of these characteristics but no other species has all of them. For example, some birds and otters use stones to cracks eggs or oysters. The New Caledonian crow is a skilled stick user. Hunting for meat is the rule for carnivores and birds of prey. Cooperation in hunting has been documented in a few populations of lions, hyenas and wolves. None of these species, however, possesses all these abilities together as in chimpanzees and humans. Therefore, behavioural similarities between these species echo their genetic and morphological proximity.

Similarities do not mean identity and many important differences exist between chimpanzees and humans. However, we first need to be clear about the similarities if we want to have sensible discussions about differences. In addition, the differences make sense only in relation to the similarities. In my opinion, we need to emphasize the similarities as these are not only widely underestimated but are often ignored. Second, much work remains to be done before we can compile a valid list of similarities; the fact that any new study on a hitherto unknown chimpanzee population increases the list of similarities should force us to be modest about what we really know about chimpanzees.[23] This is worrying as patience is not one of humanity's strong points and scientists are all too willing to draw hasty conclusions.

Yes, only humans speak, use computers and have aeroplanes to fly between continents. Yes, only humans have invented the wheel, symphonies and arrows. Yes, only humans write books and laws and have institutionalized bodies to impose society's rules and religious beliefs. But most of these aspects are very recent innovations in our 100,000-year-long history. Nobody would disagree that all these have only been possible because of what humans already were prior to the advent of agriculture some 8,000 years ago. Not, I think, that agriculture and, particularly, the industrial revolution have revolutionized what humans have become. However, we were foragers for many millennia and the later human 'revolution' has been contingent upon these forager abilities. It is fascinating to see the many parallels between our forager past and chimpanzees.

Humans as worldwide conquerors are very different from the chimpanzees. If humans have impressive similarities to chimpanzees, we have seen that they differ in some striking ways. Sexual competition between men has become inimical to women and in that process much of women's freedom has disappeared. At the same time, the demographic changes that led to this have increased human fertility and survival. Both have made humans a successful species able to invade many habitats and environments that have remained inaccessible to chimpanzees, who themselves are rather more successful at adapting to new ecological conditions than either gorillas or orang-utans. This ecological adaptability has selected for levels of behavioural and cognitive flexibility that are not seen in any other large-brained species.

Sex competition that is originally an individual challenge, as we see in many bird and small mammal species, has in slow-maturing species under high predation pressure, like the chimpanzees, become a social challenge. This shift from individual to social does not remain in the sexual domain, as the pursuit of reproductive success does not limit itself to mating. The acquisition of new mates, of safe social living conditions, and of secure access to food resources is essential to improve reproductive success. The acquisition of new mates in close social groups is only possible when individuals cooperate together against other social groups. Such intergroup conflicts further favour the development of cooperative actions. But shared goals do not mean identical strategies as males remained competitors. This requires special flexibility in adapting to ever-changing social circumstances in a fission–fusion society so typical to both humans and chimpanzees.

In *chimpanzees*, cooperation and social intelligence are used as tools to adapt to the interactions between these reproductive constraints and feeding competition. Leopard pressure adds another layer of complexity as it imposes further restrictions on the possibilities within the social domain. The higher the pressure, the more cohesive the individuals have to be and the higher the food competition between group members. More cooperation between particular social partners is the best way to mitigate such ecological constraints. With a risk of predatory attack of over one per year, the pressure on each individual is very important, and chimpanzees are then seen to provide direct altruistic acts towards all group members. Thereby, they directly increase survivorship for every one in the group. Altruism and cooperation under specific ecological circumstances become the best answer to make social life for the individual beneficial. However, under the high food availability of the African tropical rainforest, all adults can meet their nutritional requirements on their own, so that males cannot control females. In addition, some dominant males directly profit from the female's involvement in the social challenges. Consequently, female transfer in chimpanzees remains a female decision.

In *humans*, the situation as inherited from our ancestors was in many ways very similar to the chimpanzees, with long maternal investment in slow-maturing children making women rarely fertile and high predation pressure by large carnivores making social life so much more important for individual protection. Altruism and cooperation to react to predation pressure, as well as intergroup violence to access more sexual partners, became regular behaviours in our ancestors to make social life beneficial. Subsequently, a yet unidentified event in our past has driven us out from the relative safe and rich forest harbour into new more open habitats. Classically it has been proposed that the tectonic changes in East Africa east of the Rift Valley, whereby the forest gave way to a savannah-like type of environment, was at the origin of our ancestors adopting more human-like behaviour. We have seen that this 'Out of Savannah' model does not apply in this form to all of human evolution, especially concerning the evolution of tool use, hunting and cooperation and altruism. In addition, it does not really concur with the discovery of new human ancestors west of the Rift Valley.[24] The alternative would be that food competition in the forest had become too great, with elephants, chimpanzees, gorillas, monkeys, bushpigs and others all eating similar food types, so that human ancestors would have been forced out of the forest and had to adapt to new habitat.

Whatever the event that triggered this change, the situation in humans diverged importantly from the chimpanzees the moment they started to colonize new regions with different food availability, and which presented new challenges to which women responded by decreasing birth interval. This has forced women to rely to a level not equalled in any other primates on men for being able to successfully bring up more than one child at a time. This resulted in the emergence of a level of cooperation within human families between wife and husband as well as other close kin not seen in any other primates. This, however, came with a price to the women, as men had suddenly much more to lose from women's infidelity and at the same time had more means to control the women. To a level not equalled in any other primates, extreme violence against the women as a strategy to control the paternity of the children became widespread and has strongly limited women's freedom of movement and position in social life; dispersal is no more only the decision of the women, and their position in the society became dependent on the men's will.[25]

Since the development of agriculture and more so since the industrial revolution, modern human societies have become so large that the notions of 'within-group' and 'stranger' have lost most of their original biological values. I suggest that by looking honestly at our past and our cousins, we might come to appreciate the conditions which resulted in our propensity for both extreme violence and solidarity, and how much these attributes might have become purely destructive. I realize that many of us may feel rather upset by this emphasis on our biological roots with the suggestion that we are not always able to free ourselves from them. We are, with some justification, so very proud of being the primate with the largest brain. This is true. However, the large size of the human testis testifies to the importance of our biological past and the need to face up to the consequences it has on our daily behaviour. Our large brains should assist us in integrating the diverse heritage of our biological roots and so develop sexual and social behaviour patterns that are not to the detriment of one sex or defenceless children.

NOTES

1. There has been a lot of discussion about the causes of war, but women and resources are among the possible explanations regularly forwarded in war accounts (Manson and Wrangham 1991, Ericksen and Horton 1992, Keeley

1996, Gat 2000, Kelly 2000, Otterbein 2004). Cultural anthropologists historically have had a tendency to discount biological explanations of human
behaviour and, therefore, will favour sociological arguments even in the
face of abundant reports showing how often women and sex are reported
as a cause of war (Sahlins 1972, Ferguson 2000, Alès 2006). The fact remains
that women and resources to feed women and their children are constantly
proposed as a cause of war. Scarcity of resources has been forwarded as an
important predictor of war and winners will generally steal resources, such
as land, food and women, to such a point that an important change in sex
ratio could be documented in prehispanic South America due to the stealing
of women (Ember and Ember 1992, Kohler and Turner 2006).

2. Similar quotes could be provided that tend to be very egocentric in the sense
that the authors grant without question the highest level of abilities to
humans and only grudgingly allow any such abilities to non-human animals.
This double standard is very common in science and requires scrutiny.

3. A review of mortality in humans due to warfare can be found among others
in Keeley (1996), Knauft (1999) and Kelly (2000). Since a state of warfare is
found in almost all human societies, the list of studies detailing the destructive effects of violence can be very long. Some more examples follow. In the
Highlands of New Guinea, 28.5% of the Dani men and 2.4% of the women
died due to warfare, while in nearby Enga, 34.8% of adult men died in wars.
In the Lowland Gebusi of New Guinea, the mortality rate due to violence
was 35.2% in adult men, and 29.3% in adult women. This last exceptionally
high value for women is because the Gebusi fight mainly over sister
exchange marriages. And an example from present days in Europe: in the
early 1900s, Montenegro was shaken by much violence during which 25%
of the adults died.

4. As happens so often, archaeological evidence is not so easy to interpret and
not every one will agree (Keeley 1996, Gat 1999). However, most agree that
signs of war are evident in archaeological times. Early indications of victims
of warfare are scarce. This should not be a surprise, since humans only started
to bury their dead about 150,000 years ago. In addition, non-perishable
weapons have only been used in the last 40,000 years. Furthermore, fixed
settlements were only started in the last 14,000 years and were observed
more with farming, which began around 8,000 years ago (fortifications
around villages appeared then). Thus, without burial, non-perishable weapons and settlements, it is simply very difficult to collect signs of warfare.
Furthermore, distinguishing accidental trauma from homicidal causes can
be a challenge: 40% of Neanderthal skeletons show head injuries, but it is
impossible to determine the cause; these include a case of a man from about
50,000 years ago with stab wounds to the chest (Gat 2000).

5. Analyses of mortality due to warfare have been compared between humans
and chimpanzees, and showed that it varies in different types of human
groups (Manson and Wrangham 1991, Wrangham 1999). New data from
chimpanzees showed that only 16 out of 175 adult and adolescent males
were victims of intergroup violence.

6. Divale and Harris (1976) made this fascinating review of data from traditional
forager societies. Sadly, female infanticide remains highly prevalent in some
human societies, where millions of girls are killed or mistreated until they die
so that sex ratios are dramatically distorted: in India, when the first born is a
girl, the sex ratio for the second infant is 132 boys to 100 girls. In China, the
one child policy produced a sex ratio of first infants of 129 boys to 100 girls.
More dramatically, if the first two infants are girls, the sex ratio of the third

infant is 225 boys to 100 girls (Low 2000). This tendency has old roots, as in the nineteenth century in Huai-Pei province when there were many wars, the sex ratio of youngsters was 117 boys to 100 girls, and an estimated 80 million men had no women to marry.

7. Frank Marlowe (2005) carried out a survey of 294 traditional forager groups, foragers in the sense that they depended for less than 10% on planted food. He showed that on average 14.08–15.55% of men are polygynous and 21–24% of women have co-wives, meaning that they share a man with other women.

8. The famous study of the Yanomamö was carried out by Napoleon Chagnon (1977) and led to many discussions about the validity of his results. The main point challenged was that Chagnon adopted an evolutionary approach to human behaviour, which was still considered as heretical in some circles. However, the constantly increasing data from different human populations concur in demonstrating the influence of a man's status on his reproductive success, be this as a hunter or warrior, wealth and/or social status. An especially detailed study of the Ache has illustrated how such an advantage can be achieved in a South American population (Kaplan and Hill 1985, Hill and Hurtado 1996). Similarly, Wiessner (2002) detailed the complex benefits for the !Kung Bushmen in Botswana that result from being a good hunter. Similar trends were documented in traditional farmers of Kenya and Mali (Borgerhoff Mulder 1990, Strassmann 1997).

9. If the results of past studies were not always clear, more recent studies in the USA, which controlled for biological relationship between the children and men of different status, clearly showed a strong effect in data collected between 1989 and 2000 (Low 2000, Hopcroft 2006).

10. After reading the chapter about paternity in chimpanzees, I assume most readers expect me to report what we know about paternity in humans. Do the women of polygynous men have more children with the harem holder or not? Do hunters or rich men have more extramarital children? The problem is that such systematic genetic data do not exist in humans (Anderson 2006). Much more information is available for other animal species, but since paternity studies pose a real ethical issue, they cannot be carried out against the will of the parents. Baker and Bellis (1995) have proposed that there is about 9% of extramarriage paternity in humans but no systematic studies can confirm this yet.

11. When did such demographic innovations happen during human evolution? Since most of the information we have about our ancestors relies on bones, it is difficult to make conclusions about demographics. However, analysis of mortalities and development tend to show that early hominids, like Australopithecines, *Homo habilis* and Neanderthals, had patterns of mortality and development rates more similar to chimpanzees than to modern humans (Hill *et al.* 2001, Konigsberg and Herrmann 2006, Skinner and Wood 2006). This would indicate that both innovations were unique features of modern humans.

12. Much of what has traditionally been said about differences in the development of humans and chimpanzees has been based on data from captive individuals. Sadly it has become clear as more studies have been carried out that this is misleading as captive animals may grow more quickly. Detailed new analyses have changed this view (Robson *et al.* 2006, van Schaik *et al.* 2006). In addition, a general correlation has been found between larger brain size and slower maturation in many species and this seems a strong and systematic phenomenon in animals. Large brain size means more

sophisticated cognitive abilities, which give an important advantage in facing daily challenges.

13. The effect of male provision on the success of women is weaker in rich environments (50%), while it increases in harsher environments (70%) (Marlowe 2001, 2003b). When fishing is important, men's contribution is higher than when gathering is more important. Many other studies have detailed the important role of the presence of a father on the survival of children and their safe weaning and healthy development (Hurtado *et al.* 1992, Kaplan *et al.* 2002, Marlowe 2003a, Quinlan and Quinlan 2008).

14. Menopause is a uniquely human characteristic, the evolution of which is still fiercely debated as it has not been shown that it is beneficial for a woman to completely forgo reproduction (Hill and Hurtado 1991, Peccei 2001). What is clear is that once menopause occurs, it becomes advantageous for these women to help their daughters' reproduction and invest in their grandchildren (Hill and Hurtado 1991, Hawkes 2003, Ragsdale 2004).

15. To stress the importance of relatedness in humans, studies of family violence in Canada, England and New Zealand revealed that children living in households with one or more step-parents are 60 times more likely to be killed at the hands of those parents (usually the step-father) and 25 times more likely to suffer from child abuse than children of the same age with their biological parents. Similarly, unrelated members of a family (spouse or adoptive children) are 11 times more likely to be murdered than blood relations (Daly and Wilson 1988). In addition, many studies have shown that certainty of paternity is a key factor in determining a man's behaviour towards his wife and child (Daly *et al.* 1982).

16. Much of this information comes from the papers of Daly and Wilson (Daly *et al.* 1982, Wilson and Daly 1995).

17. The dunnock became a classic example of the advantages of different pairing systems (Davis 2000). Additional studies in swallows have confirmed the disadvantage of polygamous pairing for female birds that are dependent on additional provision to successfully bring up their offspring. Recent detailed studies carried out with the Dogon in Mali (Strassmann 2000), and the Kipsigi (Borgerhoff Mulder 1990) and Datoga of Kenya (Sellen *et al.* 2000), and others (Josephson 2002) show that men have much to gain from polygamy but this mating system is disadvantageous for women in terms of having fewer surviving children and having them grow up more poorly in the first 3 years of life.

18. Numerous accounts have been documented in the literature of humans falling victims to predators (see, for example, Gould and Yellen 1987, Hart and Sussman 2005). It has been suggested that the structure of the camps of forager groups was influenced by the risk of predation on humans (Gould and Yellen 1987).

19. The idea of the accumulation in caves of human ancestor bones by leopards that obviously had a liking for such prey is very disturbing to us (Washburn 1957, Lockwood *et al.* 2007). However, it is an exact illustration of the fact that throughout our past we have been victims of large carnivore attacks and that has moulded our social structure and behaviour.

20. Some experimental psychologists, not familiar with the importance of ecological conditions, favour exaggerated contrasts between chimpanzees and humans without taking into account the key role of ecological conditions faced during upbringing as well as the specific socio-ecological and economic conditions in which individuals live (Povinelli and Vonk 2003, Tomasello *et al.* 2005, Jensen *et al.* 2007, Herrmann *et al.* 2007). Recent psychological

studies considering the social relevance of the experiment convincingly showed greater abilities in chimpanzees totally in line with field observations (Horner *et al.* 2006, Inoue and Matsuzawa 2007, Whiten *et al.* 2007).

21. This project was a large collaborative effort, which tried to understand the cross-cultural differences in the propensity of humans to be altruistic and willing to punish to maintain altruism (Henrich *et al.* 2001, 2005, 2006, Gintis *et al.* 2003). The 'ultimate game' is presented in such a way that the identity of the participant is not known and it is played only once, so that direct social pressure cannot play a role. Nevertheless, we know that in such experimental settings, humans are much more generous than in the real world (Levitt and List 2008); therefore, we should look at these results more as a way to reveal cultural differences than as a measure of altruism in humans, as the situation is very artificial.

22. Recent observations on more chimpanzee populations have documented a surprising array of different tool techniques developed to gain access to food sources (Boesch *et al.* in press b); chimpanzees in Senegal and Uganda dig wells in the dry season and use sponges to access the water, chimpanzees in Congo dig out tubers full of water, Goualougo and Loango chimpanzees extract honey from well-protected beehives, Gombe, Goualougo and Loango chimpanzees use tools to extract grubs from termite mounds or insects and honey from underground beehives, Taï chimpanzees use hammers on a daily basis for months to eat very hard-shelled nuts. Recently, Fongoli chimpanzees in Senegal were seen to use 'spears' to kill bush-babies hiding in holes (Pruetz and Bertolani 2007).

23. The two new studies of chimpanzees in Central Africa, Loango in Gabon and Goualougo in Congo, have presented unique evidence of the ability of chimpanzees to combine the use of up to five tools in order to gain access to food resources (Boesch *et al.* in press b, Sanz and Morgan 2007). A new study on the Fongoli chimpanzees in Senegal, one of the driest and hottest environments inhabited by chimpanzees, revealed for the first time that they use caves or cool water pools to protect themselves from the heat (Pruetz 2001). Similarly, new observations on Ugalla chimpanzees in Tanzania showed that it is within their cognitive abilities to use sticks to dig out underground storage food (Hernandez-Aguila *et al.* 2007).

24. The 'Out of Savannah' model as proposed by Raymond Dart in the early 1990s has been strongly challenged both by new findings of early human ancestor fossils west of the Rift Valley, such as the *Sahelanthropus tchadensis*, *Orrorin tugenensis* in Tchad, as well as by comparisons done of so-called 'human-like' skills in the chimpanzees as explained in this book. Remains of early human ancestors have also been found east of the Rift Valley and therefore part of the idea of the Out of Savannah model might play a role there too.

25. Recent worldwide studies show clearly that violence against women is a function of the position given to women in the society; the more gender equality is seen in a society, the less that women are the only victims of violence (Archer 2006).

8

Postscript: Fédora's fate

Fossey migrated as a young 10-year-old female into the North Group of the Taï study chimpanzees during the summer months of 1988, possibly at the same time as Vénus, another immigrant. She was small but strongly built, carrying a proud pink sexual swelling and was immediately very confident with the males of the community. She had her first baby we called Diane in February 1992. Sadly, as happens so often, this first baby did not survive for very long. She disappeared within 10 days. Fossey started to have sexual swellings again within weeks and was seen to spend all her time with the adult males. On 12 November 1993, Fossey gave birth to her second daughter, Fédora. The little girl grew rapidly and was a healthy and curious chimpanzee. I remember her suckling eagerly at her mother's nipples at the same time looking at the world around her, including me, the strange biped, while her mother cracked nuts with a stone hammer on a root anvil. Fédora was very keen on nuts, which she received from her generous mother during the first 3 years. With great enthusiasm, she soon tried to crack them herself with whatever looked to her inexperienced eyes like a pounding tool, but was mostly ineffective. She used to steal intact nuts from her mother's collection and tried to pound them with her hands, a stick, a soft branch and even a piece of a broken termite mound. Whenever her mother collected more nuts and abandoned her hammer on the anvil, she would rush to use it, often even forgetting to place the nut. However, by trial and error, she quickly corrected such typical infant mistakes and, at the age of 6 years, had become an expert nut cracker, very assiduous although not yet very efficient. Her younger brother, Faust, would look at her with wide eyes.

Fossey remained a relaxed mother and was often seen alone with her two offspring feeding and resting, seeming happily at ease. Fédora was an active youngster and it was not uncommon to see her take young Faust on her belly and walk with him. When 9 years old, Fédora started to develop the first small swellings and Nino, then an adolescent male, immediately showed some interest. Life was going well and soon, a little sad, we expected Fédora to transfer to a neighbouring community.

In March 2003, I was horrified to see Fédora using only her left hand for walking, holding the right one in the air as immobile as possible. An iron cable was clearly visible; some of the bones of her hand were exposed – the bones of two fingers, showing white, were hanging like small sticks, while the other fingers simply were not there! I could not believe it. Fédora had suddenly become disabled and her survival was uncertain. New cocoa and coffee fields had been cut within the National Park in areas where the chimpanzees were used to foraging, and they continued to do so from time to time without being aware of any imminent danger. However, farmers, although illegal in the park, do not tolerate incursions into their crops and place snares all around the fields. These snares are made of iron cables sold as bicycle brake cables and animals that get caught in them are either condemned to starve to death, or when found by the poacher are killed and eaten.

When caught in a snare, chimpanzees are in complete shock because of the sudden inexplicable pain. They try immediately to free themselves by pulling at the snare with all their strength, which increases the pain as they pull harder and harder. The snare cable cuts deep into the skin within seconds. In such a situation, trapped chimpanzees scream in pain and panic, and I have seen them spin round and round, their whole body on the ground, trying to free themselves from the snare. Eventually twisted to the limit, the cable breaks and the chimpanzee quietens down, freed mainly from the sensation of being trapped, and possibly the pain stops getting worse. The soft skin and flesh of young Fédora provided little resistance to this treatment and she literally peeled and ripped her whole hand. She was visibly in pain and moved very slowly, not letting anyone come close to her and whimpered at any attempt by her mother to tend her wound. As a baby, she would have sought refuge on her mother's belly who then, eventually, might have been able to remove the cable. As a juvenile, however, she was still too young to know how to remove the snare while at the same time too scared by the pain to allow anyone near her. How would Fédora, such a keen nut cracker, henceforth manage with this badly crippled hand?

During the following weeks, Fédora's distress continued until, slowly, the saliva she applied regularly to the bare bones of her wrist worked miracles. It stopped bleeding and started to heal. I followed the progress of this healing process, worried about an infection, as we had lost another juvenile a long time ago because of a much less serious wound that developed gangrene. Fédora was, however, a tough cookie, and slowly became more active again, although the total absence of any sexual swelling was a clear sign of her continuing distress. After some weeks, Fédora's hand bones started to drop off – the naked bones that had still somehow held together, fell off, one by one, until the cruel human's snare cable fell off as well. Fédora's life resumed a form of normality, normality, however, with only one hand. Eventually, after a 3-year delay, she must have transferred into a new chimpanzee community, and I hope she will be spared any other human depredations.

Throughout Africa, humans protect their fields from animal incursions and chimpanzees are among the losers as well. At the same time, new fields are cleared to satisfy the needs of an ever-growing human population, with bushmeat supplying the protein requirements of these populations. Although many different animal species are killed in this way, it is again the great apes who suffer particularly severely because of their slow rates of reproduction. Chimpanzees reproduce for the first time when 14 years old and have one infant every 5 years. Such a slow rate makes them particularly vulnerable to poaching and the bushmeat crisis that is affecting all regions in Africa renders them highly threatened. In Africa, the 'empty-forest syndrome' is developing, by which forests stand intact but are empty of wildlife. The intelligence of chimpanzees cannot help them adapt to gunshots. Chimpanzee populations have plummeted from over a million 100 years ago to fewer than 200,000 some 30 years ago. Today, no one knows exactly how many are left, and where they survive.[1] Furthermore, with no forest, there are no chimpanzees. The forests are still disappearing rapidly. We have to act NOW on all levels or wild chimpanzees will shortly be gone for good.

Humans have the right to live and provide for themselves. In the tropical forest regions, however, once the forests are gone, the lives of the local human populations will no longer be the same. Simply – no forests, no rain. The already drying climate in tropical Africa will intensify and this constitutes a major challenge for humanity. In my opinion, the right of chimpanzees to live in freedom in their habitat should not be viewed as competing with human rights – they are complementary. The mutual survival of both is not to the disadvantage of either, but a guarantee of survival for both. And it is of interest to all of us – the green lungs of the tropics are the guarantee of climate stability for all of us. Therefore, we must help, in the interests of all: help the local human populations and their chimpanzee neighbours to survive in harmony in and around the tropical rainforests. The fascination with chimpanzees should be transformed into active support, either by supporting the actions of specialized organizations, like the Wild Chimpanzee Foundation (www.wildchimps.org), or by lobbying governments to provide more effective support for national parks and other protected areas as well as the sustainable management of tropical forests.

NOTES

1. It comes as a surprise that after decades of conservation projects in Africa, we are still unable to say how many chimpanzees, gorillas, bonobos or many other forest species are still present. Large amounts of money have been invested to help improve their survival but too little effort has been made to monitor how the animals fare and whether conservation projects really do profit them.

References

BOOKS ON RELATED TOPICS

Alès, C. 2006. *Yanomami: L'Ire et le Désir*. Paris: Editions Karthala.

Baker, R. and Bellis, M. 1995. *Human Sperm Competition*. London: Chapman and Hall.

Boesch, C. and Boesch-Achermann, H. 2000. *The Chimpanzees of the Taï Forest: Behavioural Ecology and Evolution*. Oxford: Oxford University Press.

Chagnon, N. 1977. *Yanomamö: The Fierce People*. New York: Holt.

de Waal, F. 1982. *Chimpanzee Politics: Power and Sex among Apes*. London: Jonathan Cape.

de Waal, F. 1989. *Peacemaking among Primates*. Cambridge: Harvard University Press.

Foley, R. 1995. *Humans before Humanity: An Evolutionary Perspective*. Oxford: Blackwell Publishers.

Goodall, J. 1970. *In the Shadow of Man*. London: Collins.

Goodall, J. 1990. *Through a Window*. New York: Houghton Press.

Hart, D. and Sussman, R. 2005. *Man the Hunted: Primates, Predators, and Human Evolution*. New York: Westview Press.

Hill, K. and Hurtado, M. 1996. *Ache Life History: The Ecology and Demography of a Foraging People*. New York: Walter de Gruyter.

Hrdy, S. 1981. *The Woman That Never Evolved*. Cambridge: Harvard University Press.

Jolly, A. 1999. *Lucy's Legacy: Sex and Intelligence in Human Evolution*. Harvard: Harvard University Press.

Keeley, L. 1996. *War Before Civilization*. New York: Oxford University Press.

Kelly, R. 2000. *Warless Societies and the Origin of War*. Ann Arbor: University of Michigan Press.

Knauft, B. 1999. *From Primitive to Postcolonial in Melanesia and Anthropology*. Ann Arbor: University of Michigan Press.

Low, B. 2000. *Why Sex Matters*. Princeton: Princeton University Press.

Morris, D. 1967. *The Naked Ape: A Zoologist's Study of the Human Animal*. New York: Dell.

Otterbein K. 2004. *How War Began*. College Station: Texas A&M University Press.

Sahlins, M. 1972. *The Use and Abuse of Biology: An Anthropological Critique of Sociobiology*. London: Tavistock Publications.

Tomasello, M. and Call, J. 1997. *Primate Cognition*. Oxford: Oxford University Press.

Wrangham, R. and Peterson, D. 1996. *Demonic Males: Apes and the Origins of Human Violence*. Boston: Houghton Mufflin Co.

HUMAN REFERENCES

Anderson, K. 2006. How well does paternity confidence match actual paternity? Evidence from worldwide nonpaternity rates. *Current Anthropology*, **47**(3): 513–520.

Archer, J. 2006. Cross-cultural differences in physical aggression between partners: a social-role analysis. *Personality and Social Psychology Review*, **10**(2): 133–153.

Arkush, E. and Stanish, C. 2005. Interpreting conflict in the ancient Andes: implications for the archaeology of warfare. *Current Anthhropology*, **46**(1): 3–17.

Bailey, R., Head, G., Jenike, M., *et al.* 1989. Hunting and gathering in tropical rain forest: is it possible? *American Anthropologist*, **91**(1): 59–82.

Boesch, C. 2007. What makes us human (*Homo sapiens*)? The challenge of cognitive cross-species comparison. *Journal of Comparative Psychology*, **121**(3): 227–240.

Borgerhoff Mulder, M. 1990. Kipsigis women's preference for wealthy men: evidence for female choice in mammals? *Behavioral Ecology and Sociobiology*, **27**: 255–264.

Chagnon, N. 1988. Life histories, blood revenge and warfare in a tribal population. *Science*, **239**: 985–992.

Cutting, A. and Dunn, J. 2006. Conversations with siblings and with friends: links between relationship quality and social understanding. *British Journal of Developmental Psychology*, **24**: 73–87.

Daly, M. and Wilson, M. 1988. Evolutionary social psychology and family homicide. *Science*, **242**: 519–524.

Daly, M., Wilson, M. and Weghorst, S. 1982. Male sexual jealousy. *Ethology and Sociobiology*, **3**: 11–27.

Descola, J. 1993. *Les Lances du Crépuscule: Relations Jivaros, Haute Amazonie*. Paris: Plon.

Divale, W. and Harris, M. 1976. Population, warfare, and the male supremacist complex. *American Anthropologist*, **78**(3): 521–538.

Ember, C. and Ember, M. 1992. Resource unpredictability, mistrust and war: a cross-cultural study. *Journal of Conflict Resolution*, **36**(2): 242–262.

Ericksen, K. P. and Horton, H. 1992. Blood feuds: cross-cultural variations in kin group vengeance. *Behavior Science Research*, **26**: 57–85.

Ferguson, B. 2000. The causes and origins of 'primitive warfare': on evolved motivations for war. *Anthropological Quarterly*, **73**(3): 159–164.

Gat, A. 1999. The pattern of fighting in simple small-scale, prestate societies. *Journal of Anthropological Research*, **55**(4): 563–583.

Gat, A. 2000. The human motivational complex: evolutionary theory and the causes of hunter-gatherer fighting. Part 1. Primary somatic and reproductive causes. *Anthropological Quarterly*, **73**(1): 20–34.

Gintis, H., Bowles, S., Boyd, R. and Fehr, E. 2003. Explaining altruistic behavior in humans. *Evolution and Human Behavior*, **24**: 153–172.

Godelier, M. 1984. *La Production de Grands Hommes: Pouvoir et Domination Masculine chez les Baruya de Nouvelle-Guinée*. Paris: Fayard.

Gould, R. and Yellen, J. 1987. Man the hunted: determinants of household spacing in desert and tropical foraging societies. *Journal of Anthropological Archaeology*, **6**: 77–103.

Hawkes, K. 2003. Grandmothers and the evolution of human longevity. *American Journal of Human Biology*, **15**: 380–400.

Henrich, J., Boyd, R., Bowles, S., *et al.* 2001. In search of *Homo economicus*: behavioral experiments in 15 small-scale societies. *American Economic Review*, **91**(2): 73–78.

Henrich, J., Boyd, R., Bowles, S., *et al.* 2005. 'Economic man' in cross-cultural perspective: behavioral experiments in 15 small-scale societies. *Behavioral and Brain Sciences*, **28**: 795–855.

Henrich, J., McElreath, R., Barr, A., *et al.* 2006. Costly punishment across human societies. *Science*, **312**: 1767–1770.

Herrmann, E., Call, J., Lloreda, M., Hare, B. and Tomasello, M. 2007. Humans have evolved specialized skills of social cognition: the cultural intelligence hypothesis. *Science*, **317**: 1360–1366.

Hill, K. 2002. Altruistic cooperation during foraging by the Ache, and the evolved human predisposition to cooperate. *Human Nature*, **13**(1): 105–128.

Hill, K. and Hurtado, A. 1991. The evolution of premature reproductive senescence and menopause in human females: an evaluation of the 'grandmother hypothesis'. *Human Nature*, **2**(4): 313–350.

Hopcroft, R. 2006. Sex, status and reproductive success in the contemporary United States. *Evolution and Human Behavior*, **27**: 104–120.

Hurtado, A., Hill, K., Kaplan, H. and Hurtado, I. 1992. Trade-offs between female food acquisition and child care among Hiwi and Ache foragers. *Human Nature*, **3**: 185–216.

Josephson, S. 2002. Does polygyny reduce fertility? *American Journal of Human Biology*, **14**: 222–232.

Kaplan, H. and Hill, K. 1985. Hunting ability and reproductive success among male Ache foragers. *Current Anthropology*, **26**(1): 131–133.

Kaplan, H., Hill, K., Lancaster, J. and Hurtado, A. 2002. A theory of human life history evolution: diet, intelligence and longevity. *Evolutionary Anthropology*, **9**: 156–185.

Kelly, R. 2005. The evolution of lethal intergroup violence. *Proceedings of the Natural Academy of Sciences of the USA*, **102**(43): 15 294–15 298.

Kohler, T. and Turner, K. 2006. Raiding for women in the pre-hispanic northern Pueblo Southwest? *Current Anthropology*, **47**(6): 1035–1045.

Konigsberg, L. and Herrmann, N. 2006. The osteological evidence for human longevity in the recent past. In *The Evolution of Human Life History* (eds. Hawkes, K. and Paine, R.), pp. 267–306. Santa Fe: School of American Research Press.

Levitt, S. and List, J. 2008. *Homo economicus* evolves. *Science*, **319**: 909–910.

Lockwood, C., Menter, C., Moggi-Cecchi, J. and Keyser, A. 2007. Extended male growth in a fossil hominin species. *Science*, **318**: 1443–1446.

Manson, J. and Wrangham, R. 1991. Intergroup aggression in chimpanzees and humans. *Current Anthropology*, **32**(4): 369–390.

Marlowe, F. 2001. Male contribution to diet and female reproductive success among foragers. *Current Anthropology*, **42**: 755–760.

Marlowe, F. 2003a. The mating system of foragers in the standard cross-cultural sample. *Cross-cultural Research*, **37**(3): 282–306.

Marlowe, F. 2003b. A critical period for provisioning by Hadza men: implications for pair bonding. *Evolution and Human Behavior*, **24**: 217–229.

Marlowe, F. 2005. Hunter-gatherers and human evolution. *Evolutionary Anthropology*, **14**: 54–67.

Mercader, J. 2002. Forest people: the role of African rainforests in human evolution and dispersal. *Evolutionary Anthropology*, **11**: 117–124.

Nowak, M. and Sigmund, K. 1992. Tit for tat in heterogeneous populations. *Nature*, **355**: 250–252.

Patterson, N., Richter, D., Gnerre, S., Lander, E. and Reich, D. 2006. Genetic evidence for complex speciation of humans and chimpanzees. *Nature*, **441**: 1103–1108.

Peccei, J. 2001. Menopause: adaptation or epiphenomenon? *Evolutionary Anthropology*, **10**: 43–57.

Perner, J., Ruffman, T. and Leekam, S. 1994. Theory of mind is contagious: you catch it from your sibs. *Child Development*, **65**: 1128–1238.

Quinlan, R. and Quinlan, M. 2008. Human lactation, pair-bonds and alloparents: a cross-cultural analysis. *Human Nature*, **19**: 87–102.

Ragsdale, G. 2004. Grandmothering in Cambridgeshire, 1770–1861. *Human Nature*, **15**(3): 301–317.

Robson, S., van Schaik, C. and Hawkes, K. 2006. The derived features of human life history. In *The Evolution of Human Life History* (eds. Hawkes, K. and Paine, R.), pp. 17–44. Santa Fe: School of American Research Press.

Sellen, D., Borgerhoff Mulder, M. and Sieff, D. 2000. Fertility, offspring quality and wealth in Datoga pastoralists: testing evolutionary models of intersexual selection. In *Adaptation and Human Behavior* (eds. Cronk, L., Chagnon, N. and Irons, W.), pp. 91–114. New York: Aldine de Gruyter.

Skinner, M. and Wood, B. 2006. The evolution of modern human life history: a paleontological perspective. In *The Evolution of Human Life History* (eds. Hawkes, K. and Paine, R.), pp. 331–364. Santa Fe: School of American Research Press.

Spoor, F., Leakey, M., Gathogo, P., *et al.* 2007. Implications of new early *Homo* fossils from Ileret, east of Lake Turkana, Kenya. *Nature*, **448**: 688–691.

Strassmann, B. 1997. Polygyny as a risk factor for child mortality among the Dogon. *Current Anthropology*, **38**(4): 688–695.

Strassmann, B. 2000. Polygyny, family structure, and child mortality: a prospective study among the Dogon of Mali. In *Adaptation and Human Behavior: An Anthropological Perspective* (eds. Cronk, L., Chagnon, N. and Irons, W.), pp. 49–67. New York: Aldine De Gruyter.

Thalmann, O., Fischer, A., Lankester, F., Pääbo, S., and Vigilant, L. 2007. The complex evolutionary history of gorillas: insights from genomic data. *Molecular Biology and Evolution*, **24**: 146–158.

Tomasello, M., Carpenter, M., Call, J., Behne, T. and Moll, H. 2005. Understanding and sharing intentions: the origins of cultural cognition. *Behavioral and Brain Sciences*, **28**: 675–691.

van Schaik, C., Barrickman, N., Bastian, M., Krakauer, E. and van Noordwijk, M. 2006. Pimate life histories and the role of brains. In *The Evolution of Human Life History* (eds. Hawkes, K. and Paine, R.), pp. 127–154. Santa Fe: School of American Research Press.

Washburn, S. 1957. Australopithecines: the hunters or the hunted? *American Anthropologist*, **59**(4): 612–614.

Wiessner, P. 2002. Hunting, healing and *hxaro* exchange: a long-term perspective on !Kung (Ju/'hoansi) large-game hunting. *Evolution and Human Behavior*, **23**: 407–436.

Wilson, M. and Daly, M. 1995. An evolutionary psychological perspective on male sexual proprietariness and violence against wives. *Violence and Victims*, **8**: 271–294.

GENERAL BIOLOGY REFERENCES

Andersson, M. 1994. *Sexual Selection*. Princeton: Princeton University Press.

Axelrod, R. and Dion, D. 1988. The further evolution of cooperation. *Science*, **242**: 1385–1390.

Axelrod, R. and Hamilton, W. D. 1981. The evolution of cooperation. *Science*, **211**: 1390–1396.

Borries, C., Launhardt, K., Epplen, C., Epplen, J. and Winkler, P. 1999. Males as infant protectors in Hanuman langurs (*Presbytis entellus*) living in multimale groups – defence pattern, paternity and sexual behaviour. *Behavioral Ecology and Sociobiology*, **46**: 350–356.

Bradley, B., Doran-Sheehy, D., Lukas, D., Boesch, C. and Vigilant, L. 2004. Dispersed male networks in western gorillas. *Current Biology*, **14**: 510–513.

Buchan, J., Alberts, S., Silk, J. and Altmann, J. 2003. True paternal care in a multi-male primate society. *Nature*, **425**: 179–181.

Clutton-Brock, T. 1991. *The Evolution of Parental Care*. Princeton: Princeton University Press.

Clutton-Brock, T. and Parker, G. 1995. Punishment in animal societies. *Nature*, **373**: 209–216.

Clutton-Brock, T., Albon, S. D. and Guiness, F. E. 1984. Maternal dominance, breeding success, and birth sex ratios in red deer. *Nature*, **308**: 358–360.

Cohas, A., Yoccos, N., Da Silva, A., Goosens, B. and Allainé, D. 2006. Extra-pair paternity in the monogamous alpine marmot (*Marmota marmota*): the roles of social setting and female mate choice. *Behavioural Ecology and Sociobiology*, **59**: 597–605.

Davis, N. 2000. Multi-male breeding groups in birds: ecological causes and social conflicts. In *Primate Males: Causes and Consequences of Variation in Group Composition* (ed. Kappeler, P.), pp. 11–20. Cambridge: Cambridge University Press.

de Waal, F. 1993. Reconciliation among primates: a review of empirical evidence and unresolved issues. In *Primate Social Conflict* (eds. Mason, W. and Mendoza, S.), pp. 111–144. New York: State University of New York Press.

Dixson, A. 1998. *Primate Sexuality*. Oxford: Oxford University Press.

Dixson, A. 2002. Sexual selection by cryptic female choice and the evolution of primate sexuality. *Evolutionary Anthropology, Supplement* **1**: 195–199.

Drea, C. and Frank, L. 2003. The social complexity of spotted hyenas. In *Animal Social Complexity: Intelligence, Culture and Individualized Societies* (eds. de Waal, F. and Tyack, P.), pp. 121–148. Cambridge: Harvard University Press.

Dugatkin, L. 1997. *Cooperation among Animals: An Evolutionary Perspective*. Oxford: Oxford University Press.

Dunbar, R. 1988. *Primate Social Systems*. New York: Cornell University Press.

Eberhart, W. 1996. *Cryptic Female Choice*. Princeton: Princeton University Press.

Edvardsson, M., Champion de Crespigny, F. and Tregenza, T. 2007. Mating behaviour: promiscuous mothers have healthier young. *Current Biology*, **17**(2): R66–R67.

Grinnell, J., Packer, C. and Pusey, A. E. 1995. Cooperation in male lions, kinship, reciprocity or mutualism? *Animal Behaviour*, **49**: 95–105.

Hamilton, W. D. 1964. The genetical theory of social behaviour (I and II). *Journal of Theoretical Biology*, **7**: 1–32.

Hoffman, J., Forcada, J., Trathan, P. and Amos, W. 2007. Female fur seals show active choice for males that are heterozygous and unrelated. *Nature*, **445**: 912–914.

Hrdy, S. 1977. *The Langur of Abou: Female and Male Strategies of Reproduction*. Cambridge: Harvard University Press.

Johnsen, A., Andersen, V., Sunding, C. and Lifjeld, J. 2000. Female bluethroats enhance offspring immunocompetence through extra-pair copulations. *Nature*, **406**: 296–298.

Komdeur, J., Daan, S., Tinbergen, J. and Mateman, C. 1997. Extreme adaptive modification in sex ratio of the Seychelles warbler's eggs. *Nature*, **385**: 522–525.

Krebs, J. R. and Davies, N. B. 1991. *An Introduction to Behavioural Ecology*. 3rd Edition. Oxford: Blackwell Scientific Publications.

Launhardt, K., Borries, C., Hardt, C., Epplen, J. and Winkler, P. 2001. Paternity analysis of alternative male reproductive routes among the langurs (*Semnopithecus entellus*) of Ramnagar. *Animal Behaviour*, **61**: 53–64.

Lefebvre, L., Nicolakakis, N. and Boire, D. (2002). Tools and brains in birds. *Behaviour*, **139**: 939–973.

Maynard Smith, J. 1982. *Evolution and the Theory of Games*. Cambridge: Cambridge University Press.

Mays, H. and Hill, G. 2004. Choosing mates: good genes versus genes that are a good fit. *Trends in Ecology and Evolution*, **19**(10): 554–559.

Moss, C. 1988. *Elephant Memories: Thirteen Years in the Life of an Elephant Family*. New York: William Morrow.

Packer, C., Scheel, D. and Pusey, A. E. 1990. Why lions form groups: food is not enough. *American Naturalist*, **136**: 1–19.

Packer, C., Gilbert, D. A., Pusey, A. E. and O'Brien, S. J. 1991. A molecular genetic analysis of kinship and cooperation in African lions. *Nature*, **351**: 562–565.

Pizzari, T. and Birkhead, T. 2000. Female feral fowl eject sperm of subdominant males. *Nature*, **405**: 787–789.

Poole, J. 1996. *Coming of Age with Elephants: A Memoir*. New York: Hyperion.

Reader, S. and Laland, K. 2002. Social intelligence, innovation and enhanced brain size in primates. *Proceedings of the National Academy of Sciences of the USA*, **99**(7): 4436–4441.

Schaller, G. B. 1972. *The Serengeti Lion*. Chicago: University of Chicago Press.

Smuts, B. and Smuts, R. 1992. Male aggression and sexual coercion of females in nonhuman primates and other mammals: evidence and theoretical implications. In *Advances in the Study of Behavior*, vol. 22 (eds. Rosenblatt J., Milinski, M. and Snowdon, C.), pp. 1–61. London: Academic Press.

Strum, S. C. 1981. Processes and products of change: Baboon predatory behavior at Gilgil, Kenya. In *Omnivorous Primates: Gathering and Hunting in Human Evolution* (eds. Harding, R. S. O. and Teleki, G.), pp. 255–302. New York: Columbia University Press.

Trivers, R. L. 1971. The evolution of reciprocal altruism. *Quarterly Review of Biology*, **46**: 35–57.

Trivers, R. L. and Willard, D. E. 1973. Natural selection of parent ability to vary the sex ratio of offspring. *Science*, **179**: 742–746.

Wolff, J. and Macdonald, D. 2004. Promiscuous females protect their offspring. *Trends in Ecology and Evolution*, **19**(3): 127–134.

Zuberbühler, K. and Jenny, D. 2002. Leopard predation and primate evolution. *Journal of Human Evolution*, **43**: 873–886.

CHIMPANZEE REFERENCES

Becquet, C., Patterson, N., Stone, A., Przeworski, M. and Reich, D. 2007. Genetic structure of chimpanzee populations. *PLOS Genetics*, **3**(4): 617–626.

Boesch, C. 1994. Cooperative hunting in wild chimpanzees. *Animal Behaviour*, **48**: 653–667.

Boesch, C. 1997. Evidence for dominant mothers investing more in sons among wild chimpanzees. *Animal Behaviour* **54**: 811–815.

Boesch, C. 2002. Cooperative hunting roles among Taï chimpanzees. *Human Nature*, **13** (1): 27–46.

Boesch, C. 2003. Is culture a golden barrier between human and chimpanzee? *Evolutionary Anthropology*, **12**: 26–32.

Boesch, C. 2005. Joint cooperative hunting among wild chimpanzees: taking natural observations seriously. *Behavioral and Brain Sciences*, **28**: 692–693.

Boesch, C. and Boesch, H. 1981. Sex differences in the use of natural hammers by wild chimpanzees: a preliminary report. *Journal of Human Evolution*, **10**: 585–593.

Boesch, C. and Boesch, H. 1983. Optimization of nut-cracking with natural hammers by wild chimpanzees. *Behaviour*, **83**: 265–286.

Boesch, C. and Boesch, H. 1984. Possible causes of sex differences in the use of natural hammers by wild chimpanzees. *Journal of Human Evolution*, **13**: 415–440.

Boesch, C. and Boesch, H. 1989. Hunting behavior of wild chimpanzees in the Taï National Park. *American Journal of Physical Anthropology*, **78**: 547–573.

Boesch, C. and Boesch, H. 1990. Tool use and tool making in wild chimpanzees. *Folia Primatologica*, **54**: 86–99.

Boesch, C., Marchesi, P., Marchesi, N., Fruth, B. and Joulian, F. 1994. Is nut cracking in wild chimpanzees a cultural behaviour? *Journal of Human Evolution*, **26**: 325–338.

Boesch, C., Crockford, C., Herbinger, I., *et al.* 2008. Intergroup conflicts among chimpanzees in Taï National Park: lethal violence and the female perspective. *American Journal of Primatology*, **70**: 1–14.

Boesch, C., Bolé, C., Nohon, G., *et al.* in press a. The role of sex, rivals and land in intergroup conflicts in wild chimpanzees. *Proceedings of the National Academy of Sciences of the USA*.

Boesch, C., Head, J. and Robbins, M. in press b. Complex toolsets for honey extraction among chimpanzees in Loango National Park, Gabon. *Journal of Human Evolution*.

Constable, J., Ashley, M., Goodall, J. and Pusey, A. 2001. Noninvasive paternity assignment in Gombe chimpanzees. *Molecular Ecology*, **10**: 1279–1300.

Crockford, C., Herbinger, I., Vigilant, L. and Boesch, C. 2004. Wild chimpanzees produce group-specific calls: a case for vocal learning? *Ethology*, **110**: 221–243.

Deschner, T., Heistermann, M., Hodges, K. and Boesch, C. 2003. Timing and probability of ovulation in relation to sex skin swelling in wild West African chimpanzees, *Pan troglodytes verus*. *Animal Behaviour*, **66**: 551–560.

Deschner, T., Heistermann, M., Hodges, K. and Boesch, C. 2004. Female sexual swelling size, timing of ovulation and male behavior in wild West African chimpanzees. *Hormones and Behavior*, **46**: 204–215.

Eddy, T., Gallup, G. and Povinelli, D. 1996. Age differences in the ability of chimpanzees to distinguish mirror-images of self from video images of others. *Journal of Comparative Psychology*, **110**(1): 38–44.

Fawcett, K. and Muhumuza, G. 2000. Death of a wild chimpanzee community member: possible outcome of intense sexual competition. *American Journal of Primatology*, **51**: 243–247.

Fisher, A., Pollack, J., Thalmann, O., Nickel, B. and Pääbo, S. 2006. Demographic history and genetic differentiation in apes. *Current Biology*, **16**: 1133–1138.

Furuichi, T. 1987. Sexual swelling, receptivity and grouping of wild pygmy chimpanzee females at Wamba, Zaire. *Primates*, **28**(3): 309–318.

Gallup, G., McClure, M., Hill, S. and Bundy, R. 1971. Capacity for self-recognition in differentially reared chimpanzees. *Psychological Record*, **21**: 69–74.

Goodall, J. 1963. Feeding behaviour of wild chimpanzees: a preliminary report. *Symposium of the Zoological Society, London*, **10**: 39–48.

Goodall, J. 1964. Tool-using and aimed throwing in a community of free-living chimpanzees. *Nature*, **201**: 1264–1266.

Goodall, J. 1986. *The Chimpanzees of Gombe: Patterns of Behavior*. Cambridge: Belknap Press of Harvard University Press.

Hernandez-Aguilar, A., Moore, J. and Pickering, T. 2007. Savanna chimpanzees use tools to harvest the underground storage organs of plants. *Proceedings of the National Academy of Sciences of the USA*, **104**(49): 19 210–19 213.

Hill, K., Boesch, C., Goodall, J., *et al.* 2001. Mortality rates among wild chimpanzees. *Journal of Human Evolution*, **40**: 437–450.

Hiraiwa-Hasegawa, M., Byrne, R., Takasaki, H. and Byrne, J. 1986. Aggression toward large carnivore by wild chimpanzees of Mahale Mountains National Park, Tanzania. *Folia Primatologica*, **47**: 8–13.

Hohmann, G. and Fruth, B. 2000. Use and function of genital contacts among female bonobos. *Animal Behaviour*, **60**: 107–120.

Horner, V., Whiten, A., Flynn, E. and de Waal, F. 2006. Faithful replication of foraging techniques along cultural transmission chains by chimpanzees and children. *Proceedings of the National Academy of Sciences of the USA*, **103**(37): 13 878–13 883.

Inoue, S. and Matsuzawa, T. 2007. Working memory of numerals in chimpanzees. *Current Biology*, **17**(23): R1004–R1005.

Jensen, K., Call, J. and Tomasello, M. 2007. Chimpanzees are vengeful but not spiteful. *Proceedings of the National Academy of Sciences of the USA*, **104**(32): 13 046–13 050.

Kano, T. 1992. *The Last Ape: Pygmy Chimpanzee Behavior and Ecology*. Stanford: Stanford University Press.

Lehmann, J. and Boesch, C. 2003. Social influences on ranging patterns among chimpanzees (*Pan troglodytes verus*) in the Taï National Park, Côte d'Ivoire. *Behavioural Ecology*, **14**(5): 642–649.

Lehmann, J. and Boesch, C. 2004. To fission or to fusion: effects of community size on wild chimpanzees (*Pan troglodytes verus*) social organisation. *Behavioral Ecology and Sociobiology*, **56**: 207–216.

Lehmann, J. and Boesch, C. 2005. Bisexually-bonded ranging in chimpanzees (*Pan troglodytes verus*). *Behavioral Ecology and Sociobiology*, **57**: 525–535.

Lehmann, J. and Boesch, C. 2009. Sociality of the dispersing sex: the nature of social bonds in West African female chimpanzees (*Pan troglodytes*). *Animal Behaviour*, **77**: 377–387.

Lehmann, J., Fickenscher, G. and Boesch, C. 2006. Kin biased investment in wild chimpanzees. *Behaviour*, **143**: 931–955.

Lukas, D., Reynolds, V., Boesch, C. and Vigilant, L. 2005. To what extent does living in a group mean living with kin? *Molecular Ecology*, **14**: 2181–2196.

Matsumoto-Oda, A. 1999. Female choice in the opportunistic mating of wild chimpanzees (*Pan troglodytes schweinfurthii*) at Mahale. *Behavioral Ecology and Sociobiology*, **46**: 258–266.

Melis, A., Hare, B. and Tomasello, M. 2006. Engineering cooperation in chimpanzees: tolerance constraints on cooperation. *Animal Behaviour*, **72**: 275–286.

Morbeck, M. and Zihlmann, A. 1989. Body size and proportions in chimpanzees, with special reference to *Pan troglodytes schweinfurthii* from Gombe National Park, Tanzania. *Primates*, **30**(3): 369–382.

Muller, M. and Wrangham, R. 2005. Testosterone and energetics in wild chimpanzees (*Pan troglodytes schweinfurthii*). *American Journal of Primatology*, **66**: 119–130.

Muller, M., Kahlenberg, S., Thompson, M. and Wrangham, R. 2007. Male coercion and the costs of promiscuous mating for female chimpanzees. *Proceedings of the Royal Society B*, **274**: 1009–1014.

Nishida, T. and Kawanaka, K. 1985. Within-group cannibalism by adult male chimpanzees. *Primates*, **26**(3): 274–284.

Nishida, T., Hiraiwa-Hasegawa, M., Hasegawa, T. and Takahata, Y. 1985. Group extinction and female transfer in wild chimpanzees in the Mahale National Park, Tanzania. *Zeitschrift für Tierpsychologie*, **67**: 284–301.

Nishida, T., Takasaki, H. and Takahata, Y. 1990. Demography and reproductive profiles. In *The Chimpanzees of the Mahale Mountains: Sexual and Life History Strategies* (ed. Nishida, T.), pp. 63–97. Tokyo: University of Tokyo Press.

Povinelli, D. and Vonk, J. 2003. Chimpanzee minds: suspiciously human? *Trends in Cognitive Sciences*, **7**(4): 157–160.

Power, M. 1991. *The Egalitarians: Human and Chimpanzee*. Cambridge: Cambridge University Press.

Pruetz, J. 2001. Uses of caves by savanna chimpanzees (*Pan troglodytes verus*) in the Tomboronkoto region of southeastern Senegal. *Pan Africa News*, **8**: 26–28.

Pruetz, J. and Bertolani, P. 2007. Savanna chimpanzees, *Pan troglodytes verus*, hunt with tools. *Current Biology*, **17**(5): 412–417.

Pusey, A., Williams, J. and Goodall, J. 1997. The influence of dominance rank on reproductive success of female chimpanzees. *Science*, **277**: 828–831.

Pusey, A., Pintea, L., Wilson, M., Kamenya, S. and Goodall, J. 2007. The contribution of long-term research at Gombe National Park to chimpanzee conservation. *Conservation Biology*, **21**(3): 623–634.

Reynolds, V. 2005. *The Chimpanzees of the Budongo Forest: Ecology, Behaviour and Conservation*. Oxford: Oxford University Press.

Rosati, A., Stevens, J., Hare, B. and Hauser, M. 2007. The evolutionary origins of human patience: temporal preferences in chimpanzees, bonobos and human adults. *Current Biology*, **17**: 1663–1668.

Sanz, C. and Morgan, D. 2007. Chimpanzee tool technology in the Goualougo Triangle, Republic of Congo. *Journal of Human Evolution*, **52**: 420–433.

Sanz, C., Morgan, D. and Gulick, S. 2004. New insights into chimpanzees, tools, and termites from the Congo Basin. *American Naturalist*, **164**(5): 567–581.

Silk, J., Brosnan, S., Vonk, J., Henrich, J. and Povinelli, D. 2005. Chimpanzee prosocial behavior. *Nature*, **437**: 1357–1359.

Stanford, C. 1998. The social behavior of chimpanzees and bonobos: empirical evidence and shifting assumptions. *Current Anthropology*, **39**(4): 399–420.

Stumpf, R. and Boesch, C. 2005. Does promiscuous mating preclude female choice? Female sexual strategies in chimpanzees (*Pan troglodytes verus*) of the Taï National Park, Côte d'Ivoire. *Behavioral Ecology and Sociobiology*, **57**: 511–524.

Stumpf, R. and Boesch, C. 2006. The efficiency of female choice in chimpanzees of the Taï forest, Côte d'Ivoire. *Behavioural Ecology and Sociobiology*, **60**: 749–765.

Surbeck, M. and Hohmann, G. 2008. Primate hunting by bonobos at LuiKotale, Salonga National Park. *Current Biology*, **18**(19): R906–R907.

Townsend, S., Slocombe, K., Thompson, M. and Zuberbuehler, K. 2007. Female-led infanticide in wild chimpanzees. *Current Biology*, **17**(10): R355–R356.

Vigilant, L., Hofreiter, M., Siedel, H. and Boesch, C. 2001. Paternity and relatedness in wild chimpanzee communities. *Proceedings of the National Academy of Sciences of the USA*, **98**: 12 890–12 895.

Wallis, J. 1997. A survey of reproductive parameters in the free-ranging chimpanzees of Gombe National Park. *Journal of Reproductive Physiology*, **109**: 297–307.

Warneken, F. Chen, F. & Tomasello, M. (2006). Cooperative activities in young children and chimpanzees. *Child Development*, **77**(3): 640–663.

Watts, D. 2004. Intracommunity coalitionary killing of an adult male chimpanzee at Ngogo, Kibale National Park, Uganda. *International Journal of Primatology*, **25**(3): 507–521.

Watts, D. and Mitani, J. 2000. Infanticide and cannibalism by male chimpanzees at Ngogo, Kibale National Park, Uganda. *Primates*, **41**(4): 357–365.

Whiten, A., Goodall, J., McGrew, W., *et al.* 1999. Cultures in chimpanzee. *Nature*, **399**: 682–685.

Whiten, A., Spiteri, A., Horner, V., *et al.* 2007. Transmission of multiple traditions within and between chimpanzee groups. *Current Biology*, **17**: 1038–1043.

Williams, J., Pusey, A., Carlis, J., Farm, B. and Goodall, J. 2002. Female competition and male territorial behaviour influence female chimpanzees' ranging patterns. *Animal Behaviour*, **63**: 347–360.

Wilson, M., Britton, N. and Franks, N. 2002. Chimpanzees and the mathematics of battle. *Proceedings of the Royal Society of London B*, **269**: 1107–1112.

Wittig, R. and Boesch, C. 2003a. 'Decision-making' in conflicts of wild chimpanzees (*Pan troglodytes*): an extension of the Relational Model. *Behavioral Ecology and Sociobiology*, **54**: 491–504.

Wittig, R. and Boesch, C. 2003b. The choice of post-conflict interactions in wild chimpanzees (*Pan troglodytes*). *Behaviour*, **140**: 1527–1559.

Wittig, R. and Boesch, C. 2005. How to repair relationships – reconciliation in wild chimpanzees (*Pan troglodytes*). *Ethology*, **111**: 736–763.

Wrangham, R. 1999. Evolution of coalitionary killing. *Yearbook of Physical Anthropology*, **42**: 1–39.

Wrangham, R., Wilson, M. and Muller, M. 2006. Comparative rates of violence in chimpanzees and humans. *Primates*, **47**: 14–26.

Index

Note. Entries can be assumed to refer to chimpanzees unless humans or another species are mentioned or implied. Most of the former relate to Taï National Park chimpanzees and are not indexed under 'Taï'. The Gombe population is also extensively discussed and not all of these references are listed under 'Gombe'. The letter 'n' in a page reference denotes a chapter endnote.

adoption, 24, 47–50
aggression
   controlling, 125–6
   feeding by humans and, 128, 135n12, 135n16
   toward females, 129, 135n10
   see also coercive male behaviour; conflict
alpha males, 26, 27, 37–9, 78
altruism
   among male chimpanzees, 47–50
   coexistence with aggression, 1, 149
   in different human societies, 151
   empathy and, 55–6, 58n10, 151
   in intercommunity violence, 50–1, 119, 128–9, 134n4
   in leopard attacks, 51–4, 134n5
   see also cooperative behaviour
anthrax, 18
ants, 64–5
apes, population sizes, 162, 163n1
   see also gorillas; humans; orangutans

baboons, 28, 30n7, 31n14, 34, 36, 40, 57n6, 70
bees, 65
   see also honey
beetles, Plate 13
birth intervals in human and chimpanzee, 145, 155
Boesch-Achermann, Hedwige (author's wife), 31n12
bonobos, 70, 130–1
Bossou, Guinea, 112, Plate 1
Boyd, William, Brazzaville Beach (1990), 81

Brazzaville Beach (1990) by William Boyd, 81
Budongo National Forest, Uganda, 30n6, 105n2, 118, 124, 127, Plate 1
Buffon, Georges-Louis, Comte de, Natural History of the Quadrupeds (1762), 111
bushmeat, 162

captive populations
   atypical behaviour in, 4–5, 131
   characteristics of, 75n8, 157n12
   studies in, 112
castration, 80, 82–3
Central African populations
   distinction from West African, 68
   limited studies of, 112–13
   tool use in, 117, 159n23
   see also Goualougo; Loango
Chimpanzee Politics (1982) by Franz de Waal, 4
chimpanzee populations
   common characteristics, 69–70
   cultural differences, 72–3, 75n7, 103–4
   diversity of, 7, 72, 131–3
   ecological differences, 4–6, 110, 116, 131
   hunting differences, 42, 109–10, 120–2
   map of locations, Plate 1
   published studies of, 2, 6, 15, 112
   significance of the Niger Gap, 68, 74n1
   tool use differences, 116–18, 124, 134nn1–2, 159nn22–3

warfare as ubiquitous among,
103–4, 127–30
see also captive populations; Central
African populations; East Africa;
West Africa
chimpanzee species
behavioural flexibility, 131–3,
136n17, 153
human knowledge of, 110–11, 132
coercive male behaviour
female choice and, 11–12, 15, 17,
30n7, 126
human polygamy and, 148
competitive behaviour, *see* food
competition; sexual competition
conflict
frequency among males, 36
group solidarity/identity and, 2
resolution, 25, 32n16, 57n3, 59n13
within-group conflicts, 78–9,
80, 119
see also aggression; intergroup
conflict
Congo, Democratic Republic of, 130,
136n15
see also Goualougo
cooperative behaviour
alongside sexual competition,
40–4, 150
claimed exclusivity to humans,
139–40
in hunting, 41–3, 119–20, 122–3
in intergroup conflict, 44–7,
103, 119
predation pressure and, 149–51, 154
rationales for, 40, 44, 58n7
see also altruism
copulation, *see* mating
cultural differences, 72–3, 75n7,
103–4

Dart, Raymond, 73, 159n24
Darwin, Charles, 12, 111
de Waal, Franz, *Chimpanzee Politics*
(1982), 4
death, attitudes to, 52, 59n12
see also killing
deforestation, 72, 135n7
demographics, 97–8, 104, 106n7, 142,
157n11
DNA, *see* genetic techniques
driver ants, 64–5
drumming, as male display, 21, 61

East Africa
absence of leopard predation,
124, 127
human origins in, 113

reason for dryness of, 68, 154
sociality of chimpanzee
populations, 75n4
see also Budongo; Gombe; Kibale;
Mahale; Uganda
ecological influences on behaviour,
4–6, 110, 114, 131, 150
see also predation pressure
emasculation, 80, 82–3
empathy, 56, 58n10, 132, 151
environment, *see* ecological influences
on behaviour
evolution, 8–9, 36, 55, 68–9, 73–4
see also human evolution
exploitative view of sex, 12
extinction, 72

farming and chimpanzees, 161
feeding by humans, 128, 135n12,
135n16
female choice
copulation refusal, 11–12
dependence on males and,
147, 155
mating with strangers, 17–18, 88–9,
96–7
paternity studies as evidence,
16, 37
and results of male conflicts, 13
selectivity in, 12, 14–15, 29n2
sociality and, 126–7
see also coercive male behaviour
female dominance in bonobos, 130
female friendships, 23–5, 124–7
female immigration
age for, 71
and fatal violence, 100
and female competition, 19, 127,
136n14
group size and, 38
female rank
emigration from natal group, 19
investment in offspring and, 25, 27,
Plate 11
production of sexual swellings
and, 19
sex of offspring and, 15, 30n5
females (chimpanzee)
altruism, in intercommunity
violence, 50–1
birth intervals in human and
chimpanzee, 145, 155
food competition by, 22, 24,
124–5, 127
genital mutilation in human, 148
imprisonment of stranger, 91–2, 94
infertility/unavailability and sex
competition, 3, 142

females (chimpanzee) (cont.)
    injury by males, 21
    involvement in intergroup conflict,
        79–80, 81–3, 94–5, 129
    male aggression toward, 129,
        135n10
    numbers of, and incursions, 98–9
    numbers of, and paternity, 37–8
    promotion of male offspring by,
        25–7, Plate 12
    social position of, 21–3, 96,
        106n11
    see also maternal investment;
        women
females (human), see women
fertility, see sexual swellings
field observation
    advantages over theory, 8–9
    difficulties in forest areas, 64, 113
    habituation to humans, 18, 91, 96,
        106n9, 128
    history of, 112–13, 132
    of humans, 143
fighting, see conflict; warfare
fission–fusion groups, 4, 70, 96,
    137n19, 153
flora, diversity of, 67
food
    dependency of human females
        and, 146
    time devoted to seeking and
        processing, 4, 124
    see also fruit; insects; meat; nut
        cracking; seeds
food competition
    by females, 22, 24, 124–5, 127
    growth of mutual support, 3
    human habitat change and, 154–5
    as a motive for warfare, 97, 101
food sharing
    with adopted infants, 48–9
    hunting roles and, 44, Plate 18
    with offspring, 115–16, Plates
        8–10
forest
    adaptations to, 9, 71, 115–19,
        121–3
    as archetype chimpanzee habitat,
        8–9, 114–15
    biodiversity of, 67
    difficulties presented to the
        scientist, 64
    human origins and, 68–9, 73–4, 154
    preservation of chimpanzees and,
        162–3
    significance of the Niger Gap, 68
    sociality of species in, 60
fossils, 68, 73, 74n2

friendships in other species, 32n15
    see also female friendships
fruit, 5, 60–2, 67, 90–1, 95

gamete sizes, 12
genetic techniques
    establishing paternity, 7–8, 13–14,
        26n1, 148
    human relationship to
        chimpanzees, 69, 111
genital mutilation in women, 148
genitalia, male
    castration, 80, 82–3
    penis length, 20, 35–6, 57n1
    sexual competition and, 35–6
    testis size, 35–6, 57n2, 145–6
gestation time, 18
glaciation influence, 68
Gombe National Park, Tanzania, 5–6,
    15, 113, Plate 1
    as atypical, 114
    female sociality in, 25, 75n4, 96,
        106n11, 124–7
    hunting techniques in, 42, 109–10,
        120–3, 135n6
    intergroup conflict, 78–9, 81, 129
    male coercion in, 15
    offspring of high-ranking females,
        27, 32n17
    as reference population, 6, 113–14
    support and cooperation in, 118,
        136n13
    tool use in, 116, 118, 134n1, 159n22
Goodall, Jane
    Gombe chimpanzee studies, 15, 81,
        84, 112, 127–30, 135n12
    as protégé of Louis Leakey, 113
    In the Shadow of Man (1970), 5, 109
    Through a Window (1995), 84
gorillas, 111, 115, 133
    human origins and, 68–9, 74n3
    sexual competition in, 34–6, 83, 90
    social structures, 70, 129
Goualougo Triangle, Congo, Plate 1
    female sociality in, 75n4, 106n11
    studies in, 112, 132–3, 134n3
    tool use, 117, 159n22–3
    see also Central African populations;
        Congo
grooming behaviour, 62–3, 125–6,
    Plate 5, Plate 7
group behaviour
    behavioural flexibility, 6
    hunting in groups, 41–3, 120–3
    intensity of conflict, 103
    as protection against predators, 59n11
    sociality and the forest
        environment, 60

solidarity/identity and conflict, 2
  see also cooperative behaviour;
    intergroup conflict; neighbouring
    groups
group cohesion
  by drumming and vocalisation, 63
  fission–fusion groups, 4, 70, 96,
    137n19, 153
  see also within-group solidarity
groups
  female departure from natal
    groups, 19, 22, 71
  multimate, and male genitalia, 36
  size of, and behaviour, 88, 106n7
  size of, and female friendships,
    24–5, 125
  size of, and paternity, 37–8, 102–3

habitats
  behavioural diversity and, 72, 114, 126
  female sociality and, 125
  human colonisation of, 146, 153–4
  see also ecological influences; forest;
    open habitats
harem species, 14, 34–5, 83
Homo species, *see* human evolution
honey, 65–6, 116–17
hormones, *see* ovulation; testosterone
    levels
Hrdy, Sarah Blaffer
  studies of Indian langurs, 32n18
  The Woman That Never Evolved
    (1981), 15
human behaviour
  anthropological studies, 143
  claims of unique behaviour, 139–40
  limits on similarities with
    chimpanzees, 9, 145, 152
  parallels in chimpanzees, 1–2, 6, 77,
    102–5, 114, 132, 152–5
human evolution
  chimpanzee and bonobo studies
    and, 130–4
  forest origins of modern humans,
    68–9, 73–4, 124, 137n19, 154
  infant development in early man,
    157n11
  probable predation pressure in,
    103, 149–50
  recent discoveries, 137n21, 154,
    159n24
  the savannah model, 113–14, 124,
    131, 154
  warfare in early man, 141
humans
  concealment of ovulation, 20, 31n10
  genetic separation from
    chimpanzees, 69, 74n3, 110–11

habituation to, 18, 91, 96, 106n9, 128
impatience, compared with
    chimpanzees, 132, 136n18
population pressure on
    chimpanzees, 161, 162
significance of male genitalia, 36
warfare among, 1, 84, 104, 105n5,
    138, 141
hunting
  by bonobos, 130–1
  collaborative hunts, 41–3, 119–23,
    Plates 14–18
  empathy and, 56
  frequency, 42, 58n9
  importance, with tool use, 133
  prey animals, 41, 58n9
  skill and age, 41, 49, 56, 59n15
  skill and human sexual success, 144
  as a species characteristic, 114
  techniques, 41–3, 48–9, 109–10,
    119–24, 135n6
hunting roles, 43–4
hyenas, 56, 59n14, 102, 149

In the Shadow of Man (1970) by Jane
    Goodall, 5, 109
inbreeding and incest, 70–1, 75n6
infanticide by females, 127, 135n12
infanticide by males
  in intergroup conflict, 89–90,
    106n8
  in other species, 28, 32n18, 116, 126
  strategies for preventing, 17,
    31n9, 126
infanticide in human societies, 142,
    147, 158n15
infants
  adoption of, 24, 47–50
  development of, 71, 145–6, 158n13
  forbearance toward stranger, 92, 94
  slow maturation of, 142, 145
  see also offspring
injuries
  due to snares, 161–2
  tending by group members, 52,
    118, 128
  from within-group conflict, 78–9,
    105n2
insects as food, 64–5, 114–18, Plate 13
intergroup conflict
  cooperative behaviour in, 44–7,
    103, 119
  degree of organisation, 83–4, 92–3,
    100–1, 107n13
  differing levels of support, 119, 128,
    136n13
  human, causes of, 138–9
  infanticide in, 89–90, 106n8

intergroup conflict (cont.)
  patrols into neighbouring
    territories, 44–7, 85–7
  prediction of attacks, 98, 107n13
  use of lethal violence, 79, 81–2,
    103–4, 105n3, 128
  see also warfare
investment in offspring, see maternal
  investment; paternity

Jolly, Alison, Lucy's Legacy: Sex and
  Intelligence in Human Evolution
  (1999), 13, 15

Kanyawara, Uganda, 30n6–7, 118,
  124–6, Plate 1
  see also Kibale National Park
Kibale National Park, Uganda, 15, 25,
  30n6–7, 120–2
  see also Kanyawara; Ngogo
killing
  death rates from human warfare,
    140, 142, 156n3, 156n5
  effect on females, 100, 142
  in intergroup conflicts, 79, 81–2,
    89–90, 103–4, 128, 156n5
  of wives by husbands, 147
  see also infanticide
kin recognition, 57n5, 58n10,
  118, 128
kissing, 91

langur monkeys, 28, 32n18, 83,
  90, 126
Leakey, Louis, 113–14
leopards, 22–3, Plate 6
  altruism when facing, 51–3, 119
  chimpanzee attacks on, 53–4,
    133, 135n8
  predation on humans, 150
  squirrels warning of, 62
  see also predation pressure
life expectancy, 70–1
Linné, Carl von, 111
lions
  cooperation in hunting, 40, 58n8,
    123–4, 152
  fights for the control of prides, 83
  infanticide in, 31n9, 90, 126
  territoriality, 102
  see also predation pressure
Loango National Park, Gabon,
  117–18, 132, 134n3, 159nn22–3,
  Plate 1
  see also Central African populations
Lucy's Legacy: Sex and Intelligence in
  Human Evolution (1999) by Alison
  Jolly, 13, 15

Mahale Mountains National Park,
  Plate 1
  absence of leopard predation, 119,
    124, 135nn7–8
  as an open habitat, 72, 115
  female intergroup transfers, 75n4,
    96, 106n8, 106n10
  female sociality and choice, 125–6
  history of studies in, 8, 30n6, 112
  hunting behaviour, 42, 120–3, 135n6
  intergroup conflict, 79, 129
  maternal investment in, 27
  support and cooperation, 118–19
  tool use, 116, 118
  within-group conflict, 79, 129
male coercion, see coercive male
  behaviour
male philopatry, 28, 70
male rank
  dominance behaviour, 21, 71
  and paternity, 37
  see also alpha males
males, chimpanzee
  adoption by, 48–50
  age of adulthood, 71
  aggression and protectiveness
    toward females, 94–5, 129, 135n10
  cooperative behaviour, 40–4, 75n5,
    103, 108n17, 152
  knowledge of female physiology,
    38–9
  numbers and conflict behaviour, 88
  numbers and hunting behaviour, 122
  offspring recognition by, 39
  voice recognition among, 63–4,
    100–1, 107n15
  see also paternity
males, human
  aggression as proposed origin for
    war, 2
  contribution to feeding offspring, 146
  genital size, 36
  unequal reproductive
    opportunities, 143–5
maternal investment, 22, 55, 142
  female rank and, 25–7, Plate 11
  in humans, 142
  sexual selectivity and, 12, 15
mating
  among neighbouring groups, 17–18
  with female prisoners, 91–2, 94, 99
  initiation of, 16, 30n8, 125–6, 135n9
  interruption by the female, 17
  success, aggression and, 100,
    135n10, 139
meat
  average consumption, 42
  friendship and food competition, 24–5

importance of, 40–1
precedence in access to, 22
see also food sharing; hunting
menopause, 146–7, 158n14
monkeys
absence of empathy, 56
chimpanzees attacked by, 109–10,
120–1
langur monkeys, 28, 32n18, 83,
90, 126
as prey animals, 41, 67, 121, Plates
14–18
subspecies divergence, 68
variety in Taï forests, 67
see also hunting
monogamous behaviour and its
limits, 8, 13
Morris, Desmond, *The Naked Ape*
(1967), 140

The Naked Ape (1967), by Desmond
Morris, 140
Natural History of the Quadrupeds (1762)
by Georges Buffon, 111
neighbouring groups
females seeking sires from, 17–18,
88–9, 96–7
knowledge of, 97–9, 100–1, 102,
107n14
nonviolent interactions, 96–7
threats from, 5
see also intergroup conflict
nest building, 66–7, 71, 73, 91
nest sharing, 49–50, 66
Ngogo, Uganda, Plate 1
female status in, 106n11, 129
hunting and meat eating, 42, 120–4
lethal inter- and within-group
conflict, 81, 105n2, 106n8, 128–9
ongoing studies in, 30n6, 132
support in conflict, 119, 136n13
tool use, 118, 131
see also Kibale
Niger Gap, 68
night calls, 66
Nishida, Toshisada, 30n6, 31n9, 75n4,
105n8, 105n10, 112
see also Mahale
nut cracking
and extraction, 115, 117, 160,
Plates 8–10
Panda nuts, 49, 66

offspring
breast-feeding and weaning, 19, 22,
25, 47–8, 71, 124, 145–7
food sharing with, 115–16,
Plates 8–10

multiple partners and survival of,
29n3, 149
numbers under polygamy and
polyandry, 148
paternal investment in, 39, 57n6,
146–7
see also infants; maternal investment
open habitats, 5, 72, 114–15
orangutans, 36, 70, 133, 145–6
orphans, 47–50
ovulation, 20, 31n10, 38–9

Parinari excelsa, 115–16
paternity
adoption independent of, 49–50
confidence in, among humans, 147
confusion and infanticide, 17,
31n9, 129
correlations with male rank, 37
evidence of female choice, 16, 21
investment in offspring, 39, 57n6,
146–7
possible recognition by males, 39
techniques for establishing, 7–8,
13–14, 29n1
patrolling frequency, 85, 87
see also territoriality
penis length, 20, 35–6, 57n1
Piltdown hoax, 111
play, 39, 57n6
poaching, 161, 162
polyandry, 143, 148
polygamy, 143, 148
see also harem species
population
densities of humans and
chimpanzees, 146
pressure and war, 1, 101, 104, 106n12
trends, 162
predation pressure
effects of absence, 5, 25
female sociality and, 124
food competition, sociality and,
124, 135n7, 154
on humans, 103, 149–51
leopards in the Taï forest, 24
on monkeys, 67–8
within-group solidarity and, 2, 3,
31n14, 54–5
xenophobia arising from, 102
see also leopards; lions
primates
philopatry among, 28
self-recognition and empathy
among, 56, 59n13
sexual competition and genital size, 36
social structure, 69–70
see also apes; monkeys

red deer, 15, 30n5, 34
reproduction
  rate of, and vulnerability, 162
  rate of, and warfare, 103
  success, male rank and group size,
    37–8
  success and human status, 143–5,
    157n8–10
revenge, 93, 106n12, 139
risk taking
  in intergroup conflicts, 44, 50–1,
    83–4, 87, 95, 99, 149
  in sexual competition, 11, 34
  when facing leopards, 53–5

Sacoglottis gabonensis, 61–2, 90, 93
savannah, *see* open habitats
'the savannah model', 113–14, 124,
    131, 154
seals, 14, 34
seeds, 65
self-recognition, 56, 59n13
selflessness, *see* altruism
sex
  bonobos, 130
  gamete size and origins of, 12
  ratio effects, 98–9, 102, 142
  value placed on by different
    species, 34
  see also mating
sexual competition, 36–9
  cooperation alongside, 40–4,
    55–6
  female infanticide and, 135n11
  female infertility/unavailability
    and, 3
  in human males, 143–5
  as motivation for warfare, 97–104,
    138–9, 145
  in other species, 34
  as a social challenge, 153
  Taï chimpanzees, 33–4
sexual dimorphism, 21, 34, 145
  see also genitalia
sexual swellings, Plate 5
  manipulative behaviour by females,
    19–20
  ovulation and, 38–9
  size variation, 20
snares, 161
sociality and environment, 60, 75n4,
    124–7
sperm lifetime, 17
sperm selection and sperm
    competition, 15, 30n4, 36
squirrels, 62
Sterculia rhinopetala, 65–6
Stumpf, Rebecca, 30n8

Taï National Park, Côte d'Ivoire,
    Plates 1–4
  behavioural differences from other
    studied populations, 9, 109–10,
    116–19, 132
  forest fauna, 22, 67
  history of studies on, 8, 112
  hunting techniques in, 41–4, 48–9,
    58n9, 120–3
  paternity studies in, 16
  tool use in, 65–6, 115–16, 134n1–2,
    159n22, 160
Tanzania, *see* Gombe; Mahale
termites, 114, 117
territoriality
  awareness of boundaries, 84,
    98, 101
  female choice of partner and, 14
  of human foragers, 146
  patrols into neighbouring
    territories, 44–7, 84–7
  of Taï chimpanzees, 84, 106n6
  use of fatal violence and, 83, 102
testis size, 35–6, 57n2
testosterone levels, 36, 57n4
Through a Window (1995), by Jane
    Goodall, 84
ticks, 62–3, 72–3
tool making, 2, 114
tool sets, 117, 133
tool use
  combination with hunting, 133
  by different populations, 116–18,
    124, 134n1–2, 159n22–3
  extracting honey and larvae, 65–6,
    116–17
  fishing for insects and larvae, 65,
    114–16, Plate 13
  in humans and chimpanzees, 152
  nut cracking and extraction, 49, 66,
    115–17, 160
  use of weapons, 53, 133
'typical' behaviour, 113

Uganda, 30n6, 117–18
  see also Budongo; Kanyawara; Kibale;
    Ngogo
ultimate game, 150–1

vegetation type, 67, Plate 1
voice recognition, 63–4, 100–1,
    107n14–15

warfare
  archaeological and historical
    evidence for human, 141, 156n4
  as characteristic of chimpanzee
    populations, 127–30

as characteristic of human
societies, 1, 84, 97, 104,
105n5, 138–43
chimpanzee behaviour viewed as,
76–7, 102–3
conduct of, among chimpanzees,
84–97
death rates from human, 140, 142,
156n3, 156n5
definition and causes, 77, 105n1,
106n12, 107n16, 108n18,
155n1
lethal intergroup violence, 79–84,
103–4, 105n3, 128
parallels between human and
chimpanzee, 102–5, 141
polygamy and, 148
sex as motivation for, 97–101, 138
sexual success of warriors, 144
see also conflict
water, access to, 152, 159n22
weapons, 53, 78, 103, 108n17, 133

West Africa, 68
see also Bossou; Taï
within-group conflicts, 78–9, 80, 119
within-group solidarity
altruism and, 54–6
between-group hostility and, 3
predation pressure as driver for, 2,
54–5, 150
wolves, 102
The Woman That Never Evolved (1981) by
Sarah Hrdy, 15
women
disempowerment, 145–9, 155
genital mutilation of, 148
as objects of warfare, 139
as victims of infanticide, 142
as victims of warfare, 138, 142
violence against, 147, 155, 159n25

xenophobia, 2, 56, 102–3, 129, 150

zoos, see captive populations